SpringerBriefs in Molecular Science

More information about this series at http://www.springer.com/series/8898

Shu Seki · Tsuneaki Sakurai · Masaaki Omichi
Akinori Saeki · Daisuke Sakamaki

High-Energy Charged Particles

Their Chemistry and Use as Versatile
Tools for Nanofabrication

 Springer

Shu Seki
Kyoto University
Kyoto
Japan

Akinori Saeki
Osaka University
Suita
Japan

Tsuneaki Sakurai
Kyoto University
Kyoto
Japan

Daisuke Sakamaki
Kyoto University
Kyoto
Japan

Masaaki Omichi
Osaka University
Suita
Japan

ISSN 2191-5407   ISSN 2191-5415   (electronic)
SpringerBriefs in Molecular Science
ISBN 978-4-431-55683-1   ISBN 978-4-431-55684-8   (eBook)
DOI 10.1007/978-4-431-55684-8

Library of Congress Control Number: 2015945125

Springer Tokyo Heidelberg New York Dordrecht London

Printed on acid-free paper

Springer Japan KK is part of Springer Science+Business Media (www.springer.com)

Preface

Ionizing radiations have played extremely important roles in the early development of quantum mechanics and chemistry. Without the discovery and subsequent development of the science of radiations, the principal aspects of the structure of atoms and molecules could not have been established. For more than a century since humans first realized the use of atomic energy, many greatly beneficial applications of radiations have emerged. The landscape of the development of material science, energy sources, or even medical devices has increasingly relied on radiations. Such widespreading uses inevitably demand better understanding of the effects of radiations. For instance, in the event of nuclear disasters and radioactive incidents, the fear of the effects of ionizing radiation on the human body and the organic media could be overwhelming. A comprehensive and quantitative understanding of the radiation chemical processes is the key element not only to overcome such fear but also to keep the effects in a beneficial regime such as is used in cancer radiotherapy, for instance.

Throughout the history of quantum mechanics, ionizing radiation has always been at the center of the experiments to understand the nature of electromagnetic waves, electrons, and atoms. Since the discovery of ionizing radiations by Roentgen, observations of the "interactions" with matter have provided numerous breakthroughs in the field of physics and chemistry and oftentimes even in medical science. By calling the radiations "X-rays", "α-rays", "β-rays", "γ-rays", etc. by their plural forms, we imply the transmitting and invisible natures of radiations, rendering them too difficult to count, visualize, and distinguish by each "ray". Only until recently, due to the new technological advancements from the last century, the damages to materials caused by a single "ray" have never been utilized as a platform for radiation detection, let alone as a technique for carving nanostructures out of materials through subsequent chemical etching. Herein, this book introduces the challenges to materials fabrication by a "single ray", once thought to be absurd, starting from the basic concept of interactions of a "single ray" with matter to the requirements for the materials and the ray in order to realize the fabrication. This book also guides potential readers from the theory to the reality of the concept: "materials nanofabrication by a particle", and hence the changing paradigm from "ionizing radiations as a source of material damages" to "ionizing radiations as a versatile tool for materials fabrication".

Acknowledgments

The authors' greatest debts in the technical explorations described in this book are to their nearest associates and graduate students. Kensaku Maeda and Satoshi Tsukuda initially discovered the developed nanostructures produced by single particles in their graduate research in the 1990s. The reactions as well as their uniqueness of non-homogeneous crosslinking in polymer materials have since been progressively discussed with Prof. Kenkichi Ishigure, Prof. Yosuke Katsumura, Prof. Hisaaki Kudo, Prof. Seiichi Tagawa, and Prof. Hiromi Shibata at The University of Tokyo. The theoretical aspects of the intra-track reactions were concreted through the discussions with Prof. Jay A. LaVern at Notre Dame University and Prof. J.M. Warman at Delft University of Technology.

In this century, Dr. Masaki Sugimoto and Dr. Yuichi Saito at the Japan Atomic Research Agency (JAEA) was one of our closest collaborators whose primary experience was shooting particles to embody the intra-track reactions yielding straightaway nanomaterials. Many scientists in JAEA (Dr. Tadao Seguchi, Dr. Yosuke Morita, Dr. Shigeru Tanaka, Dr. Masahito Yoshikawa, Dr. Tomihiro Kamiya, Dr. Yasunari Maekawa, Dr. Atsuya Chiba, and Dr. Kazumasa Narumi) have discussed with us and continuously encouraged the projects on single-particle nanofabrication.

Many of the interesting and fascinating nanomaterials have been discovered by Mr. Shogo Watanabe, Dr. Atsushi Asano, Ms. Hiromi Marui, Mr. Hoi-Lok Cheng, Mr. Wookjin Choi, Mr. Tuchinda Wasin, Mr. Michael T. Tang, Mr. Yuta Maeyoshi, Mr. Shotaro Suwa, Mr. Yuki Takeshita and Mr. Akifumi Horio. The exciting discoveries on the novel functional 1-D nanomaterials have been supported and guided by essential contributions from two excellent scientists: Dr. Kazuyuki Enomoto and Dr. Katsuyoshi Takano.

The single-particle nanofabrication technique (SPNT) has recently expanded into international research projects with the Inter-University Accelerator Centre at New Delhi, India, and some collaborative works with the group have been subsumed into the new concept of single-particle linear polymerization (STLiP) of small organic molecular systems, which is also one of the major subjects in this book.

Contents

Chapter 1
High-Energy Charged Particle Interaction with Matter

1.1 Preliminary Remarks

The terminology "quantum beam" was initially defined in Japan toward the end of the past century to represent a comprehensive set of radiations, photons, and charged particles whose energy exceeds the energy of visible light. Since the discovery of "ionizing radiations" by Röntgen, the term "stream of quanta" was defined by the interaction of a stream with matter. Ionizing radiation, for instance, is a set of ionization interactions between a stream of particles and matter. To date, based on the development of the coherent stream of Quanta and the precise analytical techniques of phenomena caused by radiation and photon interaction with matter, not only conventional "ionizing radiations" but also photons have shown to cause "ionization" reactions, leading to a strong motivation to redefine the "stream" with the term "quantum". In order to unify such a concept to various radiations, we consider accelerated charged particles with sufficiently high kinetic energy compared with that of electrons in matter as the "quantum", and discuss the interactions of the "quantum" in matter, especially organic and polymeric materials in terms of the specific chemical reactions induced by them.

Radiation chemistry has a long history of utilizing precise and sophisticated theoretical and experimental aspects for the analysis of primary, secondary, and subsequent radiation-induced phenomena in materials. At this point, the focus is on the physicochemical interactions and reactions in organic materials, which interplay at a few picoseconds after the primary physical interaction processes. The incident charged particle (e_i) in organic materials releases its kinetic energy via elastic and inelastic collisions into the "sea of electrons" of materials, setting the primary inelastic collision processes in motion:

$$e_i + AB \rightarrow AB^+ + e_i + e_s \tag{1.1}$$

$$e_i + AB \rightarrow AB^{n+} + e_i + ne_s \tag{1.2}$$

© The Author(s) 2015
S. Seki et al., *High-Energy Charged Particles*,
SpringerBriefs in Molecular Science, DOI 10.1007/978-4-431-55684-8_1

$$e_i + AB \rightarrow AB^{**} + e_i \quad \rightarrow \quad AB^+ + e_i + e_s \tag{1.3}$$

$$e_i + AB \rightarrow AB^{+*} + e_i + e_s \quad \rightarrow \quad \text{misc.} \tag{1.4}$$

$$e_i + AB \rightarrow A^+ + B + e_i + e_s \tag{1.5}$$

$$e_i + AB \rightarrow A^+ + B^- + e_i \tag{1.6}$$

$$e_i + AB \rightarrow \quad AB^* + e_i \tag{1.7}$$

$$e_i + AB \rightarrow A + B + e_i \tag{1.8}$$

$$e_i + AB \rightarrow AB^{+*} + e_i + e_s \tag{1.9}$$

$$e_i + AB \rightarrow AB^- \tag{1.10}$$

$$e_i + AB \rightarrow A^- + B \tag{1.11}$$

Here, AB and e_s are the organic molecules in the target materials and secondary electrons, respectively. Equations (1.1) through (1.6) are the ionization processes of molecule AB; however, under the standard thermodynamic conditions of the molecules, the processes represented by Eqs. (1.2)–(1.6) are the minorities and therefore negligible. For the other interaction processes, the relative energy deposition through the respective processes depends strongly on the atomic composition, chemical structures, and phases of the target materials as well as on the energy of the incident charged particle. The dominant processes in the condensed phase of the organic materials also shift dramatically in time after the incident. The minimum energy of an incident charged particle is set at around 100 eV, where the dominant process of interaction turns into elastic collision events in the energy range lower than ~100 eV, leading to direct energy conversion into thermal energy of the target materials [1].

The quantitative contribution of the processes given as Eqs. (1.1) through (1.11) can be discussed in terms of a "cross section" of each interaction. The value of cross section is the probability of the interaction of an incident particle with matter. For instance, photons as an analog of charged particle are also absorbed (interacted) by the molecules with the probability of the molar extinction coefficient whose unit is $mol^{-1} \, dm^3 \, cm^{-1}$, and even in the present case, the overall unit dimension is equal to m^2, which is the cross section equivalent to an area unit. The interaction represented by Eq. (1.7), giving directly the excited states of a molecule, is formally identical to the interaction of the photons in UV–vis range, and actually provides an approximate of optical absorption cross sections. This is called an "optical approximation," indicating the precise analysis of the energy (momentum) loss spectroscopy of an incident charged particle gives an estimate of optical absorptivity of the molecules in the energy range [2]. The dependence of the cross section as a function of the energy of incident charged particle is

Fig. 1.1 A variety of
energy dissipation processes
depending on the energy of
incident charged particles
in the gas-phase water
molecules

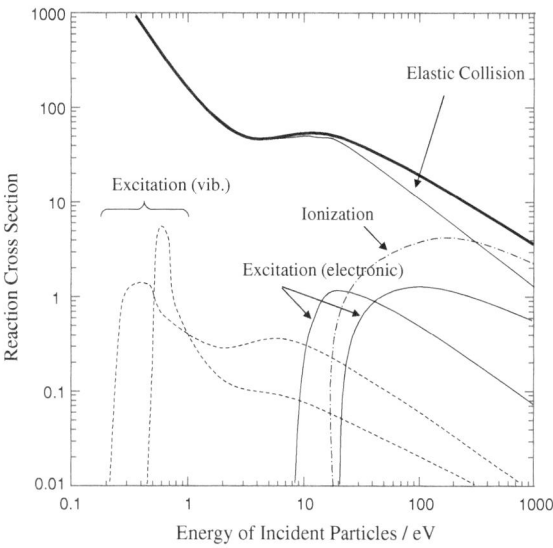

illustrated in Fig. 1.1. With increase in the incident energy of a charged particle, ionization cross section rises dramatically from the energy of a few tens of eV, and overcomes the inelastic collision cross section in the range over 100 eV. Thus the charged particle accelerated up to MeV order decelerates via the interactions with ionization events of the target molecules represented by Eqs. (1.1) through (1.6), typically by Eq. (1.1). In the ionization events, the required energy to produce one ion pair has been of interest and is characterized by a material-dependent W-value. For instance, the typical W values were reported as ~30 eV and 20 eV in water at the gas and liquid phases, respectively. Based on the optical approximation, the lower threshold of the energy loss of an incident charged particle was typically observed for the process of Eq. (1.7) giving directly the excited states of the target molecules in comparison with the threshold set for Eqs. (1.1) through (1.7), and then we may conclude the total number of events in the energy loss is major for the processes given by Eq. (1.7). However, we should also notice that the total cross section of the process is far higher for the elastic collision than those of all the other inelastic collision processes of the equations, thus eventually the absolute number of excited states are directly produced by the kinetic energy transfer from an incident charged particle to the target molecules. For instance, the W value for an incident electron with the kinetic energy of 30 eV was measured as W ~40 eV, suggesting that the major part of the electron will not cause ionization events along the thermalization (deceleration and slowing down) processes. The released energy from the electrons is, in this case, converted directly the thermal energy by elastic collision events in the media.

In practical materials, especially in their condensed phases, the events characterized by Eqs. (1.1) through (1.7) may often be independent with non-Markov property, whereas the events and processes are principally stochastic. Unlike the processes observed in gas phases, this leads to the actual events that are inter-played in the condensed media, leading to a variety of a set of relative yield of the transient species depending on the target organic molecules. Especially in the case of chemical reaction induced by high-energy charged particles, apparently we take care of the major contributions from ionic transient species to the sub-sequent chemical reactions. Almost all the processes in Eqs. (1.1) through (1.11) are initiated via Coulomb interactions and/or direct momentum transfer processes, thus the ionic intermediates are often produced within a few ~100 ps after passing the incident charged particle through the media. In the condensed media, the ion recombination through geminate and/or non-geminate recombination processes often produce excited states of the molecules whose yield is considerably higher than that of the excited state given through the direct processes of (1.7) and (1.3). The processes represented by Eqs. (1.8) through (1.11) are minor pathways but might be important in some specific molecular systems depending on the chemi-cal structures, for instance, Eqs. (1.10) and (1.11) reflect the major pathways in condensed molecules with sufficiently high electron affinity at a few tens of ps after the initial interactions of incident charged particles [3]. The yield of the tran-sient species given in the equations are often discussed quantitatively in terms of "G-value" (number of events or species produced by 100 eV absorbed by the media), and has been precisely measured and accumulated as it is one of the most important parameters in the field of radiation chemistry [4].

1.2 Interaction of Radiations in Materials

Radiation can penetrate a large variety of materials without any easy way to stop them. Powerful penetrating power is one of the most important characteristics of radiations for application in radiography. Simultaneously, high transmittance in the medium also secures homogeneous spatial distribution of the species produced via the interactions, allowing us to determine the G-values with sufficiently high accuracy. This is the most important superiority of radiation chemical processes in comparison with those produced by photochemical reactions which are initiated by the energy-selective transition of some specific molecules, and hence, nonhomoge-neous distribution of the primary excited states and subsequent reactive intermedi-ates. The homogeneity (uniformity) of reactions has been the key for analysis of the reactions induced by conventional γ-rays, X-rays, and electron beams (β-rays and cathode-rays); in contrast, a variety of new ionizing radiations have also been developed and used based on high-energy photons and charged particles. Especially, the high-energy charged particles can be focused into a small area, leading to facile "nanosized" ionization and excitation of medium. The nonhomo-geneous processes may spoil the quantitative analysis of the reactions, however, the processes induced by high-energy charged particles have been developed as unique tools for cancer radiotherapy [5] and nanomaterial fabrication in recent years. The fundamental aspects of the interactions are introduced in the following section.

1.3 Stopping Power of Materials for High-Energy Charged Particles

Radiation-induced reactions and their subsequent effects on the medium are often referred to as "radiation damages" of the materials. The terminology is usually misleading due to the impression that relatively larger net energy is deposited by the incident ionizing radiation than those of photo- and thermochemical reactions. In fact, although each ionization and excitation event given via the interaction of radiations possesses far higher energy per event, the events are distributed sparsely with large spatial intervals as long as ~100 nm, hence the net energy deposited by radiation is almost equivalent or lower than normal, low enough not to cause a significant rise in temperatures of the target materials. For instance, we used the unit of "Gy" to account for the deposited energy by radiations, which was defined as the energy per mass (J kg^{-1}). With typical γ-rays source, it takes over a few hours to irradiate a target with a dose of 100 kGy, and even with the assumption of the energy given to the target at an instance and no thermal diffusion, the temperature rises up to only 20°. Such nature allows us to presume that the energy is solely transferred to the materials and to discuss the mechanism of radiation chemical processes without any virtual regard to the macroscopic thermal effects by irradiation of conventional ionizing radiations such as γ- X-, and/or β-rays. The single event is spatially discrete. Ultimately, since the energy of an event is approximately around 100 eV, it has been a common consensus that the single "ray" cannot produce any materials itself.

The common consensus was overturned by use of high-energy "heavy" charged particles such as heavy ions accelerated up to keV–MeV ranges. The statistical and theoretical modeling of the interactions of high-energy charged particles had been well developed and is represented as the following sophisticated formula of Bethe in terms of the stopping Power (S) of materials with atomic number of Z_t for incident charged particle with the atomic number of Z_i, [6, 7]

$$S = \frac{2\pi m_p N Z_t Z_i^2 e^4}{m_e E} \ln\left(\frac{4m_e E}{m_p I}\right) \qquad (1.12)$$

$$S = \frac{2\pi N Z_t e^4}{E} \ln\left(\frac{E}{I}\right) \qquad (1.13)$$

where e, N, E, m_e, and m_p are elementary charge, number of atoms per unit volume of the target materials, energy of the incident particle, the mass of electron, and the mass of protons, respectively. The value of I was defined as an averaged excitation energy of materials, given by the average of binging energy of electrons in the target materials (electrons for ionization). In the present case, we primarily choose organic molecule/polymer systems as the target, thus we found the relatively lower value of I as 30–50 eV, and generally, the approximation of $I = 10 Z_t$ eV provides better approximation for a variety of target materials. Negligibly, small effects of the excitation of inner shell electrons have also been reported for organic materials, however, simultaneously some revision of the formulae are required in materials with high atomic number of atoms and for incident charged particles with extremely high velocity. Equations (1.12) and (1.13) are valid for heavy charged particles (atomic number: Z_i) and electron irradiations, respectively. The simplified expression of the form (1.13) was given as a result of collision event considered between electrons (electrons in the target materials and an incident electron). The logarithmic part of the equations suggest clearly that both the equations are appreciable for the incident particle with enough high velocity (momentum) than orbital velocity of electron in the target materials, because no stopping power is derived from $E = I$. This applicability of the equations is widely known as Bohr's condition. It should be noted that these equations are no longer valid for the incident particle with significantly high velocity relative to the light speed where the Lorentz transformation is required for the mass and velocity of the particles.

The simplified Eqs. (1.12) and (1.13) were established based on Born approximation where (i) elastic collision has small contribution to the total energy loss processes of an incident particle, (ii) the processes giving excited states of the target molecules were fully stochastic and the total cross section of the states were represented as the sum of oscillator strength. Finally, the changes in momentum of the incident particle are also negligible before and after the energy deposition events. To get more details on the discussion on the conditions and approximations, please refer the articles [8].

The criterion for the effective range of the incident particle velocity is given simply as follows [9]:

$$v \gg v_0 \approx \frac{e^2}{\hbar} Z_2^{2/3}. \tag{1.14}$$

Equation (1.14) is a so-called Massey's criterion. In case of hydrogen atoms in the target materials, the velocity of electrons is equivalent ($v = {\sim}v_0$) to that of accelerated charged particles with the energy of 25 keV per nuclei, suggesting the coverage of Eqs. (1.12) and (1.13) as the particle faster than the velocity. The extension of the classical equations has also been of interest, along with a variety of extended models proposed by Lindhard [1], Firsov [10], etc. Mostly in these approaches, the semi-empirical formulation for the extension of I were examined, giving more universal formula with the wider criterion toward the lower limit of the velocity of incident charged particles.

Dependence of stopping power in Eqs. (1.12) and (1.13) on E, the differentials of both equations are given as

$$\frac{dS}{dE} = \frac{2\pi m_p N Z_t Z_i^2 e^4}{m_e} \frac{1}{E^2} \left\{ 1 - \ln\left(\frac{4m_e E}{m_p I}\right) \right\}, \tag{1.15}$$

$$\frac{dS}{dE} = 2\pi NZ_t e^4 \frac{1}{E^2}\left\{1 - \ln\left(\frac{E}{I}\right)\right\}, \tag{1.16}$$

which apparently show respective maxima of S at $E = 60$ keV and 140 eV, respectively, by assuming $dS/dE = 0$ and $I = 50$ eV for proton particle (1.12) and electrons (1.13). This implies that the energy is far higher than the maximum energy of electrons used for microscopy and lithography techniques (a few tens~hundreds eV for SEM, TEM, etc.) given any further increase in energy only causes the reduction of reaction efficiency in the media. In contrast, accelerated heavy ion particles such as Ga ions are often the choice of beams in the focused ion beam systems (FIB), and the typical energy of the particle is set around a few tens of keV which is far lower than the maxima evaluated by Eq. (1.15) at E ~60 keV/nuclei. Thus, small enhancement of acceleration voltage in the FIB system will cause dramatic increase of the reaction efficiency in the target.

A maximum stopping power or the "Bragg peak" is where the maximum energy deposition is realized in density along the trajectories of the charged particle. In case of the particles bearing far higher energy than Bragg peak, the particle deposits its kinetic energy gradually as it travels into the material, with an increasing S and the slowing down processes, most of all the kinetic energy, are eventually deposited within limited spatial area at the end of the trajectory. This characteristic of the position-specific energy deposition in the material has been well analyzed and stimulated for the purpose of cancer radiotherapy with high-energy proton and heavy charged particle beams from accelerator setups. Equations (1.12) and (1.15) also imply that energy deposition schema of protons with sufficiently high energy and the other light charged particles are almost identical to those of electrons represented by Eqs. (1.13) and (1.16).

With regard to the incident particles, the energy "releasing" processes are discussed quantitatively with the value of Linear Energy Transfer (LET) that is almost identical to the Stopping Power (S). The value of LET is defined as the energy released by the incident charged particle along the unit length of the trajectories, and in the case of bending and winding of the trajectory by the scattering processes of the particle, the value was defined along the trajectories. However, the stopping power of S is defined as the energy loss of the particle per unit thickness of the target, leading to inconsistency for the charged particle with the lower kinetic energies. For the radiation chemical processes, the reactions are distributed spatially along the trajectory, and the value of LET provides better understanding of the processes.

For instance, the value of LET of conventional radiation sources such as γ-rays is 0.2–0.3 eV nm^{-1} in water and organic systems, and the interactions of γ-rays are not continuous but discrete along the trajectories. Assuming the interaction between a photon in the γ-rays and a molecule as a Compton scattering, the energy transfer from the photon to electron is 50–100 eV in one collision event. This also estimates the interval of the collision events as long as a few hundred nanometers away from each other from the value of LET as 0.2–0.3 eV nm^{-1}. According to Eq. (1.13), the value of LET is also given as 0.2–0.5 eV nm^{-1} for electrons with the kinetic energy of 100 keV ~1 MeV, suggesting almost identical primary processes to γ irradiation. In contrast, approximately 10,000-fold higher value of LET was derived from Eq. (1.12) for 400 MeV Kr particles [11], leading to the shorter spatial interval of collision events overlapping each other. The overlap is principally uniaxial along the trajectory, and eventually produces nanometer-sized cylindrical area with highly condensed reactive intermediates. This is defined as an "ion track" (particle track) as illustrated in Fig. 1.2. The energy deposited by the passing charged particle in the ion track seems to be extremely high in comparison to the averaged energy given by photo- and thermochemical reactions, however, the energy density is not enough to realize total conversion of organic molecules in the area. For instance, the value of LET of 400 MeV Kr particle reaching up to 5000 eV nm^{-1} (this value is extraordinary high in the radiation chemical processes) produces the cylindrical nonhomogeneous reaction field whose radial size is equal to 5 nm. If benzene molecules are condensed into the small disk-shaped area of 5 nm radius, 1 nm thick, the total number of the molecules will be around 600, and the deposited energy by the particle is equivalent to 8 eV per molecule, if distributed equally (Fig. 1.3). Of course, an extremely high energy is given at the center of an ion track, but the density distribution is the reciprocal function of the distance from the trajectory center. Thus the molecules at the outer part (>a few nanometers from the center) of an ion track only receive identical or smaller energy to the case of photoexcitation (typically a few eV).

The approaches to realize immobilization of organic molecules presented in this area via radiation-induced reactions are the following: (i) a use of chain reactions initiated by radiation chemical processes, and propagated to lead to condensation of the molecules, (ii) use of polymer molecules as the target materials, and "nanogel"

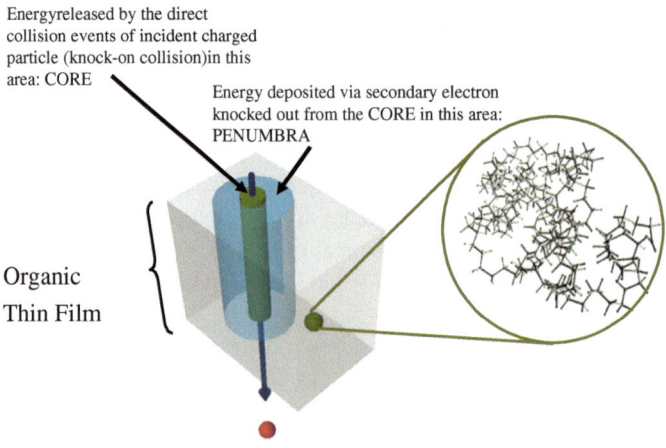

Fig. 1.2 Schematic illustration of an ion track

Fig. 1.3 Sizes and concept of intra-track reactions for nanofabrication

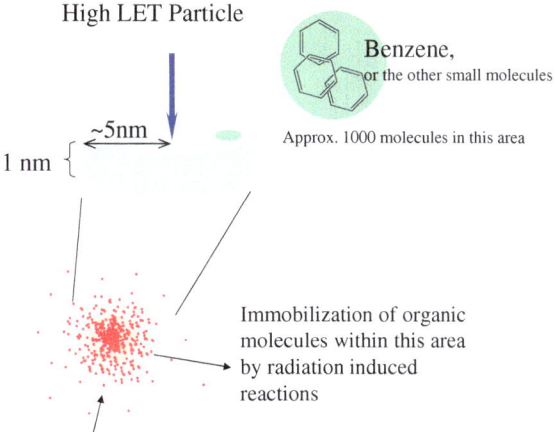

High LET Particle

Benzene, or the other small molecules

Approx. 1000 molecules in this area

~5nm

1 nm

Immobilization of organic molecules within this area by radiation induced reactions

Radial distribution of reactive intermediates

induced via cross-linking reactions by radiation chemical processes. Especially the latter has been effective for immobilization and nanomaterials fabrication because only a few crosslinks are required for the total immobilization of macromolecular chain (gelation). With use of macromolecule with molecular weight of 10^5 Dalton, the effective volume of the molecules equals ~100 nm^3 (at the density of the polymer of 1 g cm^{-3}), suggesting only "one" crosslink is necessary for gelation of the polymer presented in the "disk (~100 nm^3)" of Fig. 1.3. It is also obvious that the size of the immobilized area depends strongly on the radial distribution of reactive intermediate to promote cross-linking reactions or initiation of the polymerization reactions. When we employ polymer materials as the targets, it is also plausible that the size and shape of the macromolecules determine the size of the area. The simple concept of organic molecule immobilization in an ion track via radiation chemical processes suggests the motivation not only to simple nanofabrication by "a" particle but also to facile control of the size by the versatile parameters of LET, reaction efficiency, size and shape of the target molecules, etc. The following chapters introduce the realization of the cylinder-like linear (1 dimensional) nanostructures with diameter of a few nanometers where conventional nanofabrication techniques, such as photo- and radiation lithography, have never reached. One particle (atom, quantum) is always a particle, which never splits into fragments unlike conventional particle beams that are a body of focused quantum particles and intrinsically scattered by interactions. The simple indivisible single particle allows us to produce one-dimensional nanostructures with ultra-high aspect ratio that have never been realistic with the conventional beam-based technique.

References

1. J. Lindhard, M. Scharff, Phys. Rev. **124**, 128 (1961)
2. M. Inokuti, Rev. Mod. Phys. **43**, 297 (1971)
3. L.G. Christophorou, D.L. McCorkle, A.A. Chiristodoulides, *Electron-Molecular Interactions and Their Applications*, vol. 1 (Academic Press, New York, 1984)
4. Y. Tabata et al., Eds., *Handbook of Radiation Chemistry* (CRC Press, Boca Raton, 1991)
5. R.R. Wilson, As the first report on the idea of cancer radiotherapy. Radiology **47**, 487 (1946)
6. H.A. Bethe, Ann. Phys. **5**, 325 (1930)
7. H.A. Bethe, Z. Phys. **76**, 293 (1932)
8. N.F. Mott, H.S.W. Massey, *The Theory of Atomic Collisions*, 3rd edn. (Oxford University Press, Oxford, 1965)
9. As a textbook, G. R. Freeman, "Kinetics of Nonhomogeneous Processes", 1987, Wiley, New York, Chapter 4, p. 171
10. O.B. Firsov, Sov. Phys. JETP **9**, 1076 (1959)
11. J.F. Ziegler, J.P. Biersack, U. Littmark, *The Stopping and Range of Ions in Solids*. Calculation Code: TRIM (SRIM) 2003 (Pergamon Press, New York, 2003)

Chapter 2
Chemistry of High-Energy Charged Particles: Radiations and Polymers

The combination of polymers and the high-energy charged particles with sufficiently high LET is the promising candidate for the feasibility of the concept: "A particle producing a material." The radiation sensitivity of polymers has been widely and extensively discussed in the field of radiation chemistry, and here we focus onto the phenomena upon irradiation to high-energy charged particles. The radiation-induced reactions in polymers generally depend on the nature of radiation sources, especially the value of LET [1–3], because of the secondary reactions among reactive intermediates produced by radiations. Among a variety of polymeric materials, Si backbone polymers; polysilanes are the simple but interesting motifs exhibiting drastic shift of the major reactions from main chain scission to cross-linking with an increase in the value of LET [4, 5]. Here, the first example of the incarnation of the concept is discussed with this polymer motif.

The cross-linking reactions in the polymeric systems are often promoted by the coupling reactions between neutral reactive intermediates, and the polysilane systems demonstrated effective formation of silyl radicals with relatively high stability to the carbon centered radical analogs [6, 7]. The coupling reactions seemed to occur within an ion track, leading to give an insoluble "nanogel" along each corresponding particle, and produce wire-like 1D-nanostructures via isolation of "nanogel" on a substrate by removing soluble uncross-linked parts [8–16].

The first demonstration is the simple three step protocol: coating of the polymers onto Si substrates at 0.2–1.0 μm thick, irradiation to a variety of MeV order high-energy charged particle from several accelerators in vacuum chambers, and washing the film in solvents to remove uncross-linked parts of the film [7, 15, 16]. After the washing and drying procedures, surface morphology of the substrate with the "nanogels" was observed directly by an atomic force microscope (AFM).

Striking contrast to the highly sensitive main chain scission reaction of polysilanes upon exposure to UV light, X-rays, and high-energy charged particle irradiation of films was found to cause gelation of the polymers for all particles, all

© The Author(s) 2015
S. Seki et al., *High-Energy Charged Particles*,
SpringerBriefs in Molecular Science, DOI 10.1007/978-4-431-55684-8_2

Fig. 2.1 Sensitivity curves (gel evolution curves) for PMPS with various molecular weights ($M_w = 68$–33 (PMPS-1), 20–16 (PMPS-2), 3.6–3.0 (PMPS-3), 1.5–1.2 (PMPS-4), 0.71–0.60 (PMPS-5) $\times 10^4$ g mol^{-1}) under irradiation with a 2 MeV ^{4}He$^+$ particles. Reprinted with permission from Seki et al. 2005. ©2005, American Chemical Society (Ref. [14])

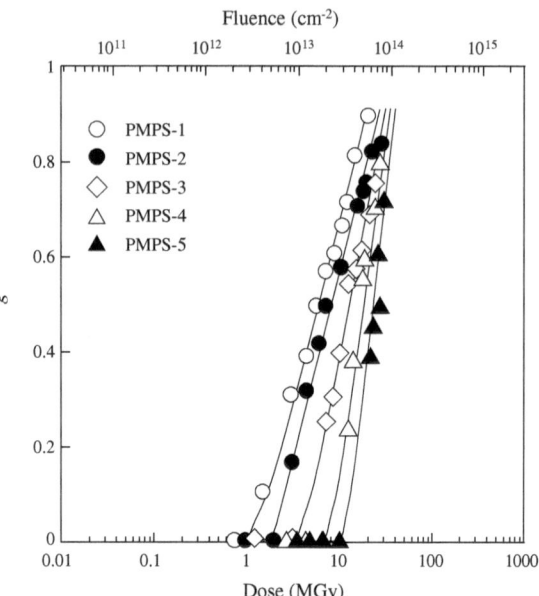

energies, and all molecular weights of poly(methylphenylsilane) (PMPS). Since the first synthesis of soluble polysilane derivatives, the effective backbone fragmentation via excitation of the σ-bonds in their main chains had been motivated to use the derivatives as a candidate for sensitive resist materials in the lithography processes. The results of high-energy charged particle irradiation are reversed completely, and this is the case of the radical coupling reaction in the extraordinary high density of the intermediate in an ion track. Macroscopic gelation behavior of PMPS is represented by the sensitivity (gelation) curves as shown in Fig. 2.1; total gel volume evolution with the irradiation of 2 MeV He$^+$ particles with the value of LET as 220 eV nm^{-1}.

In Fig. 2.1, the gel fraction increased dramatically with the absorbed dose (D), reaching up to 1 where the entire polymer film becomes insoluble against any kind of solvents. Yields of radiation-induced reactions have been expressed as G-values, where G is the number of molecules changed (produced or damaged) per 100 eV energy absorbed by the media. This quantification of the chemical yields were already introduced by the initial analysis of the reactions in the polymer materials by Charlesby [17] and the characteristic yields of cross-linking and main chain scission are given by $G(x)$ and $G(s)$, respectively. According to the statistical theory of the reactions, the values of $G(x)$ and $G(s)$ can be estimated by the following Charlesby–Pinner relationship [18]

$$s + s^{1/2} = \frac{1}{q}\left(p + m\,M_{n,0}D\right) \tag{2.1}$$

$$s = 1 - g \tag{2.2}$$

Table 2.1 $G(x)$ measured for PMPS with a variety of molecular weight upon irradiation to high-energy charged particles[a]

Polymers	Molecular weight $(/10^4 \text{ kg mol}^{-1})$	$G(x)$				
		2 MeV H 15 eV/nm	2 MeV He 220 eV/nm	0.5 MeV C 410 eV/nm	2 MeV C 720 eV/nm	2 MeV N 790 eV/nm
PMPS-1	68–33	0.0018 $(0.0021)^b$	0.0049 $(0.0052)^b$	0.021 $(0.022)^b$	0.072 $(0.0079)^b$	0.082 $(0.0095)^b$
PMPS-2	20–16	0.0021	0.0095	0.052	0.081	0.15
PMPS-3	3.6–3.0	0.0030	0.019	0.07	0.18	0.21
PMPS-4	1.5–1.2	0.0075	0.021	0.075	0.20	0.26
PMPS-5	0.71–0.60	0.019 $(0.021)^b$	0.061 $(0.089)^b$	0.18 $(0.19)^b$	0.27 $(0.33)^b$	0.34 $(0.42)^b$

[a]All the data are quoted from Refs. [3, 8, 14]
[b]Values in the parenthesis were estimated by Eqs. (2.5) and (2.6)

$$G(x) = 4.8 \times 10^3 q \tag{2.3}$$

$$G(s) = 9.6 \times 10^3 p \tag{2.4}$$

where p and q are the probability of scission and cross-linking, s and g are sol and gel fractions traced, m is the molecular weight of a monomer unit, and $M_{n,0}$ is the number average molecular weight before irradiation. The $G(x)$ values are calculated using these equations for irradiation of PMPS-1 with 2 MeV He$^+$, H$^+$, C$^+$, and N$^+$ particles are compared in Table 2.1.

It is obvious that the values of $G(x)$ depends on the molecular weight of the polymers, especially for high LET charged particles. Besides chain length, there are no differences in the chemical structures of the polymers, thus $G(x)$ values should be identical for all polymers to a first approximation.

The Charlesby–Pinner equation of (2.1) is reduced to a function of M_n and M_w (number and weight average molecular weight) as follows:

$$\frac{1}{M_n} = \frac{1}{M_{n,0}} + \left(p - \frac{q}{2}\right)\frac{D}{m} \tag{2.5}$$

$$\frac{1}{M_w} = \frac{1}{M_{w,0}} + \left(\frac{p}{2} - q\right)\frac{D}{m} \tag{2.6}$$

where $M_{w,0}$ is the initial weight average molecular weight of the target polymer before irradiation. Poly(di-n-hexylsilane) (PDHS) is the polymer with the identical Si backbone to PMPS, and stable neutral radical formation has been observed upon irradiation to UV and γ-rays through Si–Si σ-bond homolytic cleavage [5]. The relatively long linear alkyl chains (n-hexyl) in contrast to PMPS are expected to act as the source of alkyl radicals, which is the primary choice to promote effective cross-linking reactions in conventional carbon-based polymer materials. The concerted effects of the stable silyl radicals and reactive alkyl radicals were

Fig. 2.2 Charlesby–Pinner plotting of Mn to absorbed dose for 2 MeV ^{4}He$^+$, 225 MeV ^{16}O^{7+}, 160 MeV ^{16}O^{7+}, 2 MeV ^{14}N$^+$, 175 MeV ^{40}Ar^{8+}, 220 MeV ^{12}C^{5+}, 20 MeV ^{4}He^{2+}, 2–45 MeV ^{1}H$^+$, 20–30 keV e$^-$, and ^{60}Co γ-rays. Reprinted with permission from Seki et al. 1999. ©1999, American Chemical Society (Ref. [4])

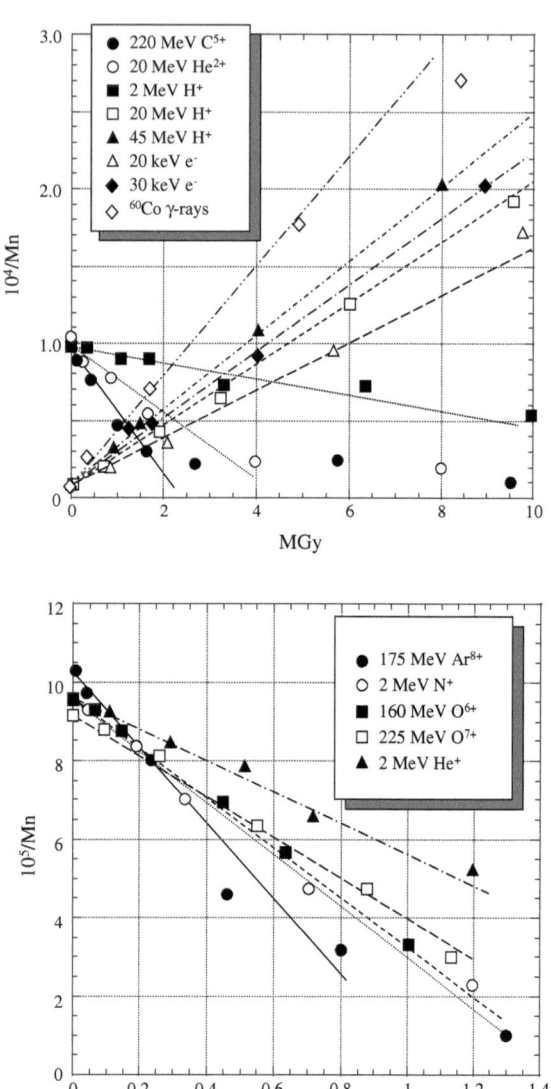

of interest to reveal the mechanisms, efficiency, and yields of cross-linking reactions depending on the density of the reactive intermediates. Here, Fig. 2.2 shows the change of M_n with an increase in the absorbed dose of a variety of radiations, suggesting the clear changes in the slope of the linear relationship. As given in Eqs. (2.5) and (2.6), the slope determines the balance of reaction sufficiency of main chain scission and cross-linking, and the inversion of the slope from positive to negative ones indicate clearly the dominant reaction caused by the radiation from main chain scission (breakdown of the polymer chain) to cross-linking (gelation). It should be noted that the change in the slope shift gradually with LET, and this is also suggestive that the coupling reaction between the silyl and alkyl radicals has been actually enhanced with the density of the species distributed linearly along the particle trajectory.

The effects of the molecular weight distribution on the radiation-induced gelation of a real polymer system were considered by Saito [19] and Inokuti [20], who traced the changes in distribution due to simultaneous reactions of main-chain scission and cross-linking analytically. However, in the present case, the molecular weight distributions of the target polymers are reasonably well controlled to be less than PDI = 1.2, and the initial distributions are predicted not to play a major role in gelation. The simultaneous change in the molecular weight distribution due to radiation-induced reactions also results in a nonlinearity of Eqs. (2.1) and (2.2). Therefore, the following equations are proposed to extend the validity of the relationship by introducing a deductive distribution function of molecular weight on the basis of an arbitrary distribution [21]:

$$s + s^{1/2} = p/q + \frac{(2 - p/q)(D_V - D_g)}{(D_V - D)} \tag{2.7}$$

$$D_v = 4\left(\frac{1}{uu_n} - \frac{1}{u_w}\right) \Big/ 3q \tag{2.8}$$

where D_g is the gelation dose, and u is the degree of polymerization (u_n and u_w are initial number-averaged and weight-averaged degree of polymerization, respectively). Equation (2.7) provides a better fit to the observed values of s at high doses than Eq. (2.1). However, the $G(x)$ derived from this fit are almost identical to those in Table 2.1, depending on the molecular weight, because the values are estimated in the low-dose region where Eq. (2.1) is sufficiently linear.

Statistical analysis of the effects of molecular weight distribution was successful partially, and well taken into accounts reaching actual yield of the chemical reaction induced not only by conventional but also by high LET radiations, in reality, high LET radiation-induced changes in molecular weight distribution was in an irrational way and opened for questions as shown in Fig. 2.3 [22]. The degradation reactions by H^+ particles (low LET particles) were traced in a typical way, with gradual shift of molecular weight distribution maxima toward the lower region and the broadenings. In contrast, the changes in the molecular weight distribution upon irradiation to high LET radiations changes exceptionally, projecting a new peak with 100-hold molecular weight range, and the initial peak was unchanged. This is suggestive that, as illustrated in Fig. 2.1, the cross-linking reactions occur only in the limited spatial area around a particle trajectory, and the polymers at the outer area remains unchanged.

During the last century, the concept of spatially limited field of chemical reactions in an ion track; "chemical core" or "chemical track" had been confirmed, visualized, and used in terms of the "damaged area" by the chemical reactions [1, 2, 23, 24]. The size of the field was limited within a few nm spaces, which

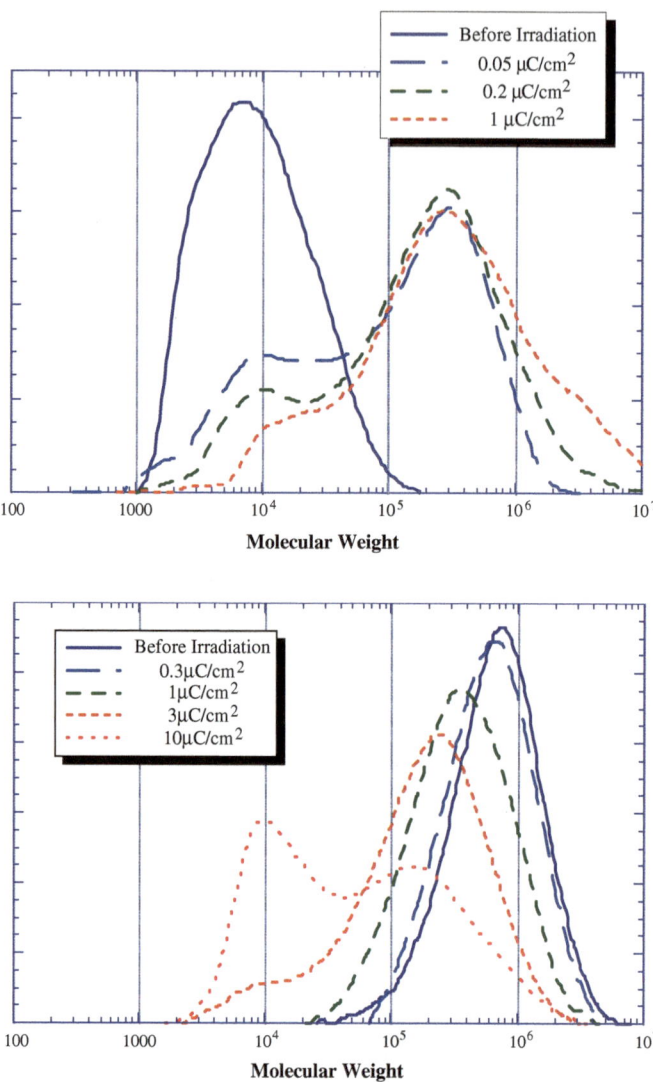

Fig. 2.3 Normalized molecular weight distribution of poly(di-*n*-hexylsilane) (PDHS) with an irradiation to 2 MeV ^{4}He$^+$ and 20 MeV H$^+$ ion beams at 295 K. Reprinted with permission from Seki et al. 1997. ©1997, Elsevier (Ref. [22])

was the best fit to the strong demands to the nanomaterials, especially nanowires, nanorods, nanothreds, nanostrings, etc., overall one-dimensional nanostructures with high aspect ratio. Motivated from the demands, the "reactions in an ion track" were designed to produce directly the materials with functions in the twenty-first century as introduced in the following sections.

References

1. J.L. Magee, A. Chatterjee, J. Phys. Chem. **84**, 3529 (1980)
2. A. Chatterjee, J.L. Magee, J. Phys. Chem. **84**, 3537 (1980)
3. S. Seki, S. Tsukuda, K. Maeda, Y. Matsui, A. Saeki, S. Tagawa, Phys. Rev. B **70**, 144203 (2004)
4. S. Seki, K. Maeda, Y. Kunimi, S. Tagawa, Y. Yoshida, H. Kudoh, M. Sugimoto, Y. Morita, T. Seguchi, T. Iwai, H. Shibata, K. Asai, K. Ishigure, J. Phys. Chem. B **103**, 3043 (1999)
5. S. Seki, H. Shibata, H. Ban, K. Ishigure, S. Tagawa, Radiat. Phys. Chem. **48**, 539 (1996)
6. S. Seki, S. Tagawa, K. Ishigure, K.R. Cromack, A.D. Trifunac, Radiat. Phys. Chem. **47**, 217 (1996)
7. K. Maeda, S. Seki, S. Tagawa, H. Shibata, Radiat. Phys. Chem. **60**, 461 (2001)
8. S. Seki, K. Maeda, S. Tagawa, H. Kudoh, M. Sugimoto, Y. Morita, H. Shibata, Adv. Mater. **13**, 1663 (2001)
9. S. Seki, S. Tsukuda, S. Tagawa, M. Sugimoto, Macromolecules **39**, 7446 (2006)
10. S. Seki, S. Tsukuda, Y. Yoshida, T. Kozawa, S. Tagawa, M. Sugimoto, S. Tanaka, Jpn. J. Appl. Phys. **43**, 4159 (2003)
11. S. Tsukuda, S. Seki, A. Saeki, T. Kozawa, S. Tagawa, M. Sugimoto, A. Idesaki, S. Tanaka, Jpn. J. Appl. Phys. **43**, 3810 (2004)
12. S. Tsukuda, S. Seki, S. Tagawa, M. Sugimoto, A. Idesaki, S. Tanaka, A. Ohshima, J. Phys. Chem. B **108**, 3407 (2004)
13. S. Tsukuda, S. Seki, M. Sugimoto, S. Tagawa, Jpn. J. Appl. Phys. **44**, 5839 (2005)
14. S. Seki, S. Tsukuda, K. Maeda, S. Tagawa, H. Shibata, M. Sugimoto, K. Jimbo, I. Hashitomi, A. Koyama, Macromolecules **38**, 10164 (2005)
15. S. Tsukuda, S. Seki, M. Sugimoto, S. Tagawa, J. Phys. Chem. B **110**, 19319 (2006)
16. S. Tsukuda, S. Seki, M. Sugimoto, S. Tagawa, Appl. Phys. Lett. **87**, 233119 (2005)
17. A. Charlesby, Proc. R. Soc. London Ser. A **222**, 60 (1954)
18. A. Charlesby, S.H. Pinner, Proc. R. Soc. London Ser. A **249**, 367 (1959)
19. O. Saito, J. Phys. Soc. Jpn. **13**, 1451 (1958)
20. M. Inokuti, J. Chem. Phys. **33**, 1607 (1960)
21. K. Olejniczak, J. Rosiak, A. Charlesby, Radiat. Phys. Chem. **37**, 499 (1991)
22. S. Seki, K. Kanzaki, Y. Kunimi, S. Tagawa, Y. Yoshida, H. Kudoh, M. Sugimoto, T. Sasuga, T. Seguchi, H. Shibata, Radiat. Phys. Chem. **50**, 423 (1997)
23. J.E. Fischer, Science **264**, 1548 (1994)
24. R.L. Fleisher, P.B. Price, R.M. Walker, *Nuclear Tracks in Solid* (University of California Press, Los Angeles, 1975)

Chapter 3
A Particle with High Energy: A Versatile Tool for Nanomaterials

The sizes of chemical cores: Leaving the arguments on inconsistency of the classical theory of stopping S given Eqs. (1.15) and (1.16) (Classical Bethe theory and Bohr theory for stopping [1–3], we will start from the equipartition theorem on energy released from a charged particle to two kinds of collision processes: nuclear collision and electronic(ionizing)collision processes. The direct momentum transfer from an incident high energy charged particles to an atom in the target causes "knock-on" collision events, and Columbic interaction between the charges of an incident particle and electrons in the target medium produces simultaneously several knocked-out secondary electrons, that is "glancing" collision events. The latter is apparently inelastic collision events, and the sum of the cross section of the interactions is estimated by the optical approximation. The equipartition theorem is applied to the above collision mechanisms, leading to the following two equations with an identical prefactor of $LET/2$, [4, 5]

$$\rho_c = \frac{LET}{2}\left[\pi r_c^2\right]^{-1} + \frac{LET}{2}\left[2\pi r_c^2 \ln\left(\frac{e^{1/2}r_p}{r_c}\right)\right]^{-1} \qquad r \le r_c \qquad (3.1)$$

$$\rho_p(r) = \frac{LET}{2}\left[2\pi r^2 \ln\left(\frac{e^{1/2}r_p}{r_c}\right)\right]^{-1} \qquad r_c < r \le r_p \qquad (3.2)$$

where ρ_c is the deposited energy density in the core area, and r_c and r_p are the radii of the core and penumbra area. The first term of Eq. (3.1) is due to the "knock-on" collision events, and the size of the area (r_c) is defined as a function of the velocity of incident charged particle. The value of r_c becomes smaller for the faster incident particle, reflecting a decrease of S in the range of E over Bragg peak, and typically $r_c \sim$ a few nm or less. Equation (3.2) is given as an inverse

© The Author(s) 2015
S. Seki et al., *High-Energy Charged Particles*,
SpringerBriefs in Molecular Science, DOI 10.1007/978-4-431-55684-8_3

square law, representing the deposited energy distribution in penumbra area, where the energy is given by the interaction between secondary electrons knocked out and the target medium. The energy spectra of the secondary electrons depend primary on the velocity of an incident particle, and the electrons dissipate their energy based on the processes given in Eqs. (1.1) through (1.11) and Fig. 1.1. The size of penumbra area (r_p) is also defined by the range of the secondary electrons, hence the maximum energy of them, and becomes typically one or three orders of magnitude larger than that of core area for MeV order charged particles.

Here, the subsequent chemical reactions after the initial energy deposition are predicted to play important roles for the target materials especially in penumbra area. Assuming cross-linking reactions as a key to immobilize the molecules in the limited area (chemical core) and produce nanogels, we can calculate easily the required density of energy to produce sufficient crosslinks for polymer gelation based on the value of $G(x)$. Generally, for gel formation in a polymer system, it is necessary to introduce one crosslink per polymer molecule. Assuming a sole contribution from the cross-linking reactions in the chemical core, ρ_{cr} is given by

$$\rho_{cr} = \frac{100\rho A}{G(x)mN} \tag{3.3}$$

where A is Avogadro's number, and N is the degree of polymerization. The value of $mN/\rho A$ reflects the volume of a polymer molecule. Substitution of $\rho_p(r)$ in Eq. (3.3) with ρ_{cr} gives the following requirement for r'. [6, 7]

$$r'^2 = \frac{LET \cdot G(x)mN}{400\pi\rho A}\left[\ln\left(\frac{e^{1/2}r_p}{r_c}\right)\right]^{-1} \tag{3.4}$$

This theoretical prediction of the size of chemical core estimates typically the value of r' as ~5 nm for $G(x)$ ~0.1 $(100\ eV)^{-1}$, M_w ~10,000 g mol^{-1}, LET ~4000 eV nm^{-1}, and r_p/r_c ~100, respectively. In addition, the sizes of chemical core are tunable not only by LET values but also by the molecular weight of the target polymers and their crosslinking reaction efficiency. The sizes have been observed in the range of 1–50 nm, which is unattainable range of the sizes by conventional nanolithography techniques particularly for fabrication of nanostructures with high aspect ratio.

A schematic of this type of non-homogeneous distribution is shown in Fig. 3.1. After the dissolution of non-crosslinked part of the polymer film, the nanogel reveals, and we can estimate the size of the corresponding "chemical core". It should be noted that Intra-molecular cross-linking, which constitutes a greater contribution to $G(x)$ in the inner region of ion tracks associated with higher densities of deposited energy, is not taken into account by the statistical treatment in Eqs. (3.1)–(3.4), and it is this case that gives rise to an underestimate of the number of crosslinks. In case of polymer films thin enough to allow the change in kinetic energy of incident particle to be negligible, a model of cylindrical energy deposition is sufficient, without needing to refer to the dependence of the radial energy distribution on the direction of the ion trajectory.

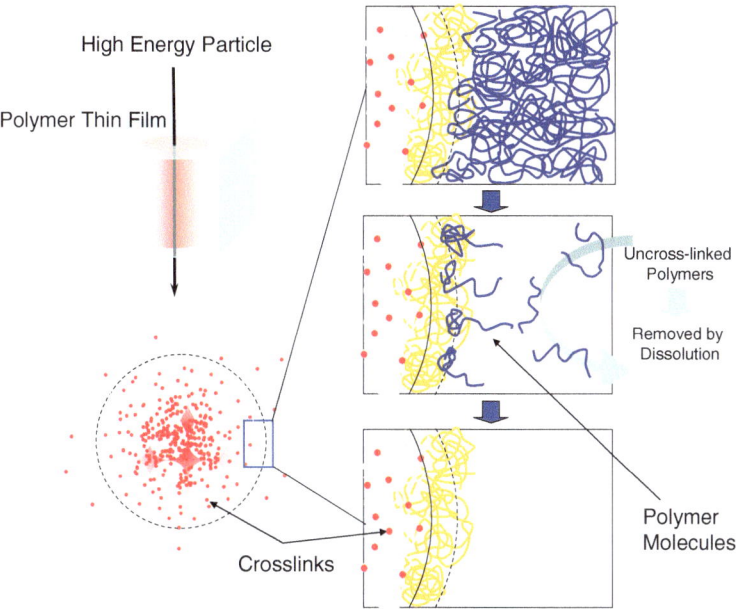

Fig. 3.1 Schematic of the non-homogeneous distribution of crosslinks in an ion track and the relative sizes of chemical core and macromolecules. Dissolution of non-crosslinked molecules after passing through an incident charged particle results in the isolation of immobilized part of polymer molecules in the target, leading to isolated "nanowires"

The determination of the size of chemical core is simply accessible by the precise trace of total gel volume fraction produced by the charge particle irradiation. The yield of gels can thus be represented with a standard sigmoid function as follows [6, 8, 9].

$$g = 1 - \exp\left[-n\pi\, r_{cc}^2\right] \qquad (3.5)$$

Here, r_{cc} is the radius of chemical core, and n represents the fluence of incident ions. The effect of the diffusion of reactive intermediates, determined using a low-energy charged particle, has been formulated as the term δr [6]:

$$r_{cc} = r' + \delta r' \qquad (3.6)$$

The value of $\delta r'$ in a typical cross-linking type polymer, poly(methylphenylsilane) (PMPS), was determined to be 0.5–0.7 nm [6]. Based on traces of the gel fraction by Eq. (3.5), the estimated values of r_{cc} are summarized for a variety of high-energy charged particles in Table 3.1.

The value of r_{cc} exceeding 1–2 nm can be traced and measured directly by AFM observation of the chemical cores with sufficiently high accuracy after their isolation onto flat substrates. For instance, AFM images of the chemical cores isolated onto Si substrate are shown in Fig. 3.2 for comparison with traces of the gel fraction.

Table 3.1 Values of r_{cc} for various ion beams observed in PMPS[a]

Ions	Energy/MeV	LET/eV nm^{-1}	r_{cc} for PMPS/nm
^{1}H^{+b}	2.0	15	0.14
^{4}He^{+b}	2.0	220	0.60
^{12}C^{+b}	0.50	410	1.1
^{12}C^{+b}	2.0	720	3.4
^{14}N^{+b}	2.0	790	3.8
^{14}N^{+c}	2.0	790	3.6
^{56}Fe^{2+c}	5.1	1550	5.5
^{28}Si^{2+c}	5.1	1620	5.9
^{28}Si^{5+c}	10.2	2150	6.1
^{40}Ar^{8+c}	175	2200	6.1
^{56}Fe^{4+c}	8.5	2250	5.8
^{56}Fe^{5+c}	10.2	2600	7.7
^{84}Kr^{20+b}	520	4100	7.9

[a]$Mn = 1.2$–1.5×10^4 g mol^{-1}
[b]Values estimated from gel traces by Eq. (3.5)
[c]Values estimated by direct AFM observation

Equation (3.5) provides a good fit for the trace of the gel fraction, and the esti-mated values of r_{cc} for both particles correspond with the values observed from the AFM micrographs. Fine patterns of chemical cores in the AFM micrographs are also given Fig. 3.2, revealing clear 1D rod-like structures (nanowires) on the substrate. It should be noted that the density of the nanowires on the substrate increased clearly with an increase in the number of the incident particles, and the observed number density of the nanowires coincided with the number density of the incident particle. This is also suggestive that "one nanowire" is produced cor-responding incident particle along its trajectory, which is a clear evidence of the model described above. The length of the nanowires is uniform in each image, and is consistent precisely with the initial thickness of the film. This is due to the geo-metrical limitation of the distribution of cross-linking reaction: the gelation occurs from the top-surface to bottom of the polymer film. Thus the length of the nanow-ires can be perfectly controlled by the present technique. Based on the measure-ment of cross-sectional trace of the nanowires by AFM, the radial distribution of crosslinks in the nanowires is discussed in terms of r_{cc} defined as the radius of the cross-section. This supports strongly the hypothesis of theoretical Eq. (3.4), where the value of r_{cc} determined primary by the molecular weight of the target polymer, the values of $G(x)$, and LET.

The concept of nm-scaled immobilization via cross-linking reaction induced high LET charged particle has been widely applied for a variety of polymeric systems via "Single-Particle Nanofabrication Technique (SPNT)". The details of the applicability are introduced in the following chapters, revealing extremely wide feasibility of SPNT to miniaturize a variety of polymer materials (not only the polymeric systems but also small organic molecular systems condensed and

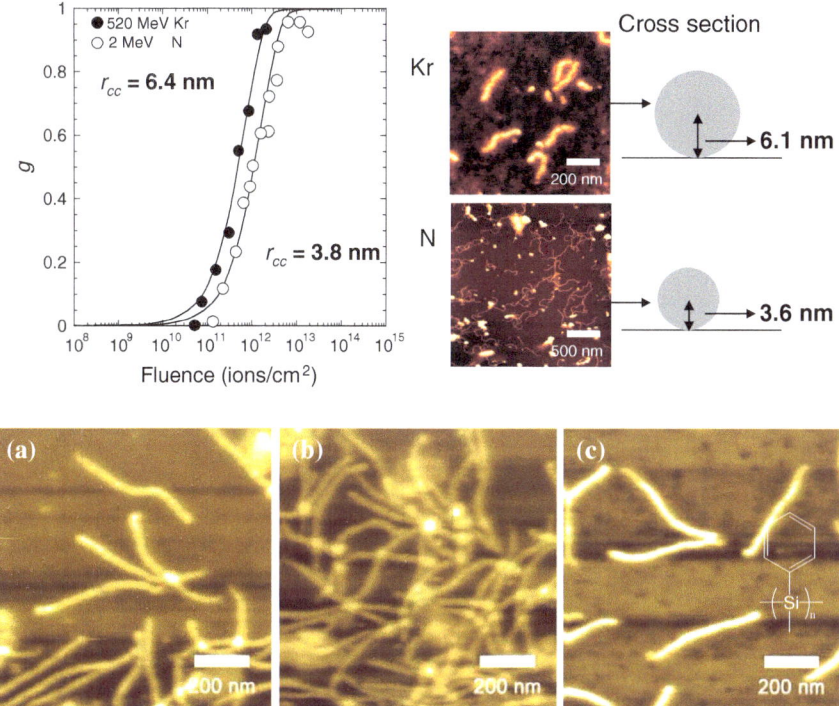

Fig. 3.2 Gel evolution curves recorded for 520 MeV ^{84}Kr irradiation of PMPS ($M_n = 0.60$–0.71×10^4 g mol^{-1}) and 2 MeV ^{14}N irradiation of PMPS ($M_n = 1.2$–1.5×10^4 g mol^{-1}). *Solid lines* denote the fit for the respective gel fraction based on Eq. (3.4). The estimated values of r_{cc} are 6.4 and 3.8 nm for the Kr and N particles. AFM micrographs were observed for the same set of polymers and particles. The fluence of Kr and N ions was set at 1.4×10^{10} and 6.4×10^9 cm^{-2}, respectively. Enlarged AFM micrographs of PMPS nanowires are also shown in (**a**)–(**c**). The nanowires were formed by 500 MeV ^{192}Au beam irradiation to (**a**, **b**) PMPS ($M_n = 1.2$–1.5×10^4 g mol^{-1}) and (**c**) PMPS ($M_n = 3.3$–6.8×10^5 g mol^{-1}) thin films at (**a**) 3.0×10^9, (**b**) 5.0×10^9, and (**c**) 1.0×10^9 particles cm^{-2}, respectively. The thickness of the target films were (**a**, **b**) 350 nm and (**c**) 250 nm. Reprinted with permission from Seki et al. 2005. ©2005, American Chemical Society. (Ref. [8])

immobilized via chain polymerization reactions). Upon extension of SPNT to a wider range of polymeric systems with a variety of backbone structures, some discrepancy was found between predicted and measured values of r_{cc} respectively by Eq. (3.4) and the direct AFM traces. For instance in case of PMPS, using the reported value of $G(x) = 0.12$ derived from radiation-induced changes in molecular weight, the values of r_{cc} calculated by Eq. (3.4) were compared with the experimental values, showing a good consistency with the experimental values for polymers with sufficient chain length ($M_n > 10^4$ g mol^{-1}). However, a considerable discrepancy occurs between the calculated and experimental results for the polymer with shorter chain lengths.

The global configuration of the polymer molecules depends heavily on the length of the polymer chains, leading to transformation from random coil (long chain) to rod-like (short chain) conformations. The gyration radius of a polymer molecule, which determines the size of a molecule spreading in the media, is correlated with this transformation. The correlation function between R_g and N is provided by the well-established Flory–Huggins theory as given in the following simple form,

$$R_g = \kappa M^\nu \tag{3.7}$$

where κ is the Flory-Huggins parameter, M is the molecular weight of the polymer, and $\nu = (\alpha + 1)/3$. Based on the persistence length of the target polymer materials, the scaling law for a helical worm-like chain model [10] results directly in an index α from Eq. (3.7). Thus, the effective volume of a polymer chain can be simply calculated as $4/3\pi R_g^3$, and the substitution of $mN/\rho A$ in Eq. (3.4) with the effective volume leads the next final expression [8, 11]

$$r'^2 = \frac{LET \cdot G(x)N^{3\alpha}}{400\pi\beta}\left[\ln\left(\frac{e^{1/2}r_p}{r_c}\right)\right]^{-1} \tag{3.8}$$

where β is the effective density parameter of the monomer unit (10^3 kg m^{-3}). Based on Eq. (3.8), the calculated value of r_{cc} is plotted again the experimental values (Fig. 3.3).

All polymers with different molecular weights, follow a single trend, and the calculated values display good correspondence in the range $r_{cc} > 7$ nm. The underestimate of r_{cc} by Eq. (3.8) for $r_{cc} < 7$ nm suggests that the initial deposition of energy and the radial dose distribution estimated by Eq. (3.2) do not account for the radial distribution of chemical intermediates and thus cannot model the concentration of cross-linking in the core of the ion track. The value of $G(x)$ increases dramatically with an increase in the density of reactive intermediates. Based on the assumption that $G(x)$ is a function of the density of deposited energy, the present results indicate that the yield of the chemical reaction is dependent on the energy density. Cross-linking reactions in ion tracks therefore have potential for not only single-particle fabrication with sub-nanometer-scale spatial resolution for any kind of cross-linking polymer, but also the study of nanoscale distributions of radial dose and chemical yield in an ion track.

The successful fit between the experimental values and Eq. (3.8) is not perfectly matched since the values of r_{cc} were determined experimentally in air by tracing the morphology of the nanowire surfaces by a scanning probe microscope, while the Eq. (3.8) was derived directly from the solution based theory of polymer backbone conformations represented by Eq. (3.7). The deviation of the measured radii from the theoretical predictions stems from the theoretically neglected nanowires-solvent interactions during the isolation of the nanowires from the non-irradiated polymer. The so-called "development" isolation was carried out by washing out of non-crosslinked polymer materials by a "rich" solvent that interacts not only to the pristine polymer but also to the nanowires themselves. The

Fig. 3.3 Correlation
between r_{cc} values
estimated by gel traces,
direct AFM measurement,
and calculations using
Eq. (3.4). Correlation
between r_{cc} values estimated
experimentally and those
calculated using Eq. (3.8).
Reprinted with permission
from Seki et al. 2005.
(Ref. [8]) ©2005, American
Chemical Society

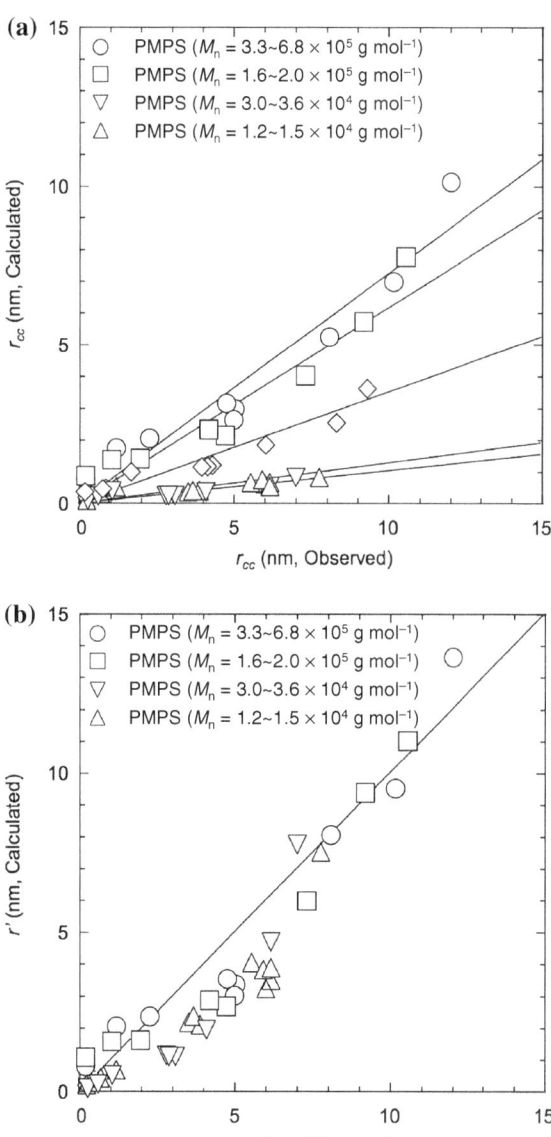

nanowires also interact strongly with the surfaces during the procedures and the
subsequent drying processes, and this is the case giving the "memorized" shape of
macromolecules in solution even at the air-substrate interfaces.

As seen in this chapter, the high energy charged particle technique, SPNT, produced one-dimensional nanomaterials with (i) ultra-fine spatial resolution equivalent of the size of a polymer molecules, (ii) extremely high aspect ratio due to the

indivisible nature of "a particle", and (iii) universal feasibility of the technique for nanofabrication. The size and dimensions of the materials produced by SPNT are in the inaccessible zone of the conventional top-down lithographic techniques so far. The wide-range of target materials for SPNT also suggests the enormous versatility of the technique to change the landscape of the molecules used in nanofabrication from only specific self-assembling organic molecules necessary for bottom-up aggregation to virtually "any" molecules with sufficient tendency for cross-linking and/or polymerization reactions.

References

1. N. Bohr, Phys. Rev. **58**, 654 (1940)
2. N. Bohr, Mat Fys Medd Dan Vid Selsk **18**, 1 (1948)
3. N. Bohr, J. Lindhard, Mat Fys Medd Dan Vid Selsk **28**, 1 (1954)
4. J.L. Magee, A. Chatterjee, J. Phys. Chem. **84**, 3529 (1980)
5. A. Chatterjee, J.L. Magee, J. Phys. Chem. **84**, 3537 (1980)
6. S. Seki, S. Tsukuda, K. Maeda, Y. Matsui, A. Saeki, S. Tagawa, Phys. Rev. B **70**, 144203 (2004)
7. S. Tsukuda, S. Seki, M. Sugimoto, S. Tagawa, Appl. Phys. Lett. **87**, 233119 (2005)
8. S. Seki, S. Tsukuda, K. Maeda, S. Tagawa, H. Shibata, M. Sugimoto, K. Jimbo, I. Hashitomi, A. Koyama, Macromolecules **38**, 10164 (2005)
9. S. Seki, K. Kanzaki, Y. Yoshida, H. Shibata, K. Asai, S. Tagawa, K. Ishigure, Jpn. J. Appl. Phys. **36**, 5361 (1997)
10. H. Yamakawa, F. Abe, Y. Einaga, Macromolecules **27**, 5704 (1994)
11. S. Seki, K. Maeda, S. Tagawa, H. Kudoh, M. Sugimoto, Y. Morita, H. Shibata, Adv. Mater. **13**, 1663 (2001)

Chapter 4
Bio-compatible Nanomaterials

4.1 Preliminary Remarks

Protein nanowires are drawing attention in various research fields. Several studies on drug delivery systems [1], scaffolds for cell adhesion [2], and diagnosis [3, 4] have been conducted using protein nanowires. Generally, self-assemblies of proteins and peptides have been employed to fabricate protein one-dimensional nanostructures because proteins and peptides are versatile building blocks for constructing well-defined one-dimensional nanostructures [5]. The diameters of the protein one-dimensional nanostructures are between 10 and 20 nm, and their lengths are up to several tens of micrometers. However, fabrication methods that take advantage of these self-assembly reactions cannot be applied to proteins and peptides which do not self-assemble. Furthermore, it is difficult to control the self-assembly reaction of protein molecules to provide nanowires of consistent length.

Recently, it was reported that the size of a polymeric particulate antigen delivery system, a nanoparticle, was important for the efficient binding and uptake of particles by antigen-presenting cells, such as macrophages and dendritic cells [6, 7]. Furthermore, the immune response changed both quantitatively and qualitatively depending on the particle size. Similarly, the specific interactions of protein nanowires with biological systems likely depend on the size of the protein nanowires. Thus, to fabricate protein nanowires that can act as a modulator, a fabrication method to ensure that it gives uniformly sized nanowires is required.

Living cells are surrounded by protein fibers, typically composed of collagen and fibronectin. These protein fibers regulate cellular responses through direct interactions with cell surface integrin receptors [8, 9]. A recent study showed that the cellular response can be controlled by laying prepared fibronectin-based layer-by-layer nanometer-sized films directly on a cell surface; the degree of cellular

© The Author(s) 2015
S. Seki et al., *High-Energy Charged Particles*,
SpringerBriefs in Molecular Science, DOI 10.1007/978-4-431-55684-8_4

response control depended on the thickness of the nanometer-sized films [10]. Thus, uniformly sized protein nanowires used as novel scaffolds should also be able to strictly control cellular responses.

SPNT is a method for fabricating one nanowire by a single charged particle whose energy can be controlled easily. It is, therefore, expected that adjusting the energy of the single charged particle will minimize damage to the structure of the protein and lead to retention of its biological activity. Here, the fabrication and characterization of size-controlled protein nanowires by SPNT [11, 12] are reviewed in this chapter, referring the usabilities by conversion to the protein nanowires that show biological activity across large areas of the nanowire, taking advantage of the avidin–biotin interaction.

4.2 Fabrication of Protein Nanowires

Human serum albumin (HSA) is the most abundant protein in human blood plasma and is a key protein involved in the biocompatibility of biomaterials. A single charged particle can fabricate one HSA nanowire along its trajectory. Non-cross-linked HSA can be removed with phosphate-buffered saline (PBS) (Fig. 4.1).

After drying, the surface of the irradiated film was observed by atomic force microscopy (AFM). The AFM images suggested that HSA nanowires with a cross-linked structure were fabricated by the ion beam (Fig. 4.2a). The diameters of the protein nanowires were about 20 nm and almost equal to the diameter of self-assembled protein and peptide nanostructures. The density of the HSA nanowires increased with increasing fluence of the charged particle (Fig. 4.2a–d); thus, by adjusting the fluence of the charged particle, the density of the HSA nanowires can be easily controlled. The irradiated HSA film showed a broad electron spin

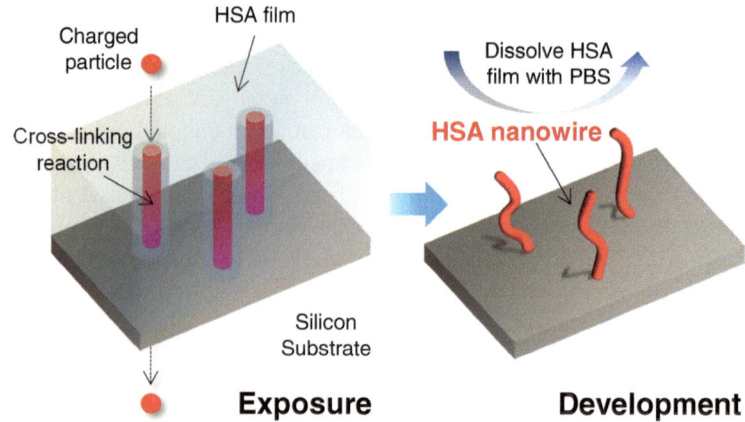

Fig. 4.1 Process for fabricating HSA nanowires by SPNT

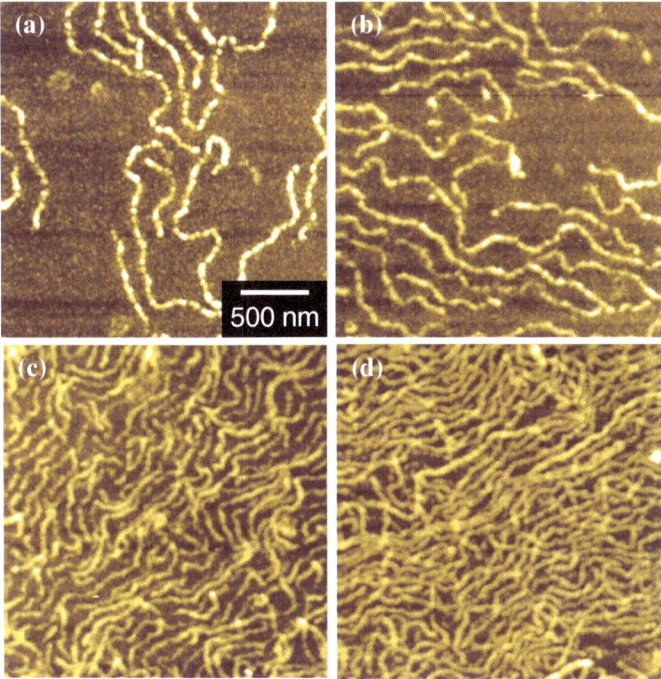

Fig. 4.2 AFM images of the HSA nanowires formed by irradiation of a HSA spin-coated film with a 320 MeV ^{102}Ru^{18+} ion beam at a fluence of **a** 1.0×10^8, **b** 3.0×10^8, **c** 7.0×10^8, and **d** 1.0×10^9 ions cm^{-2}. Reproduced with permission from Ref. [11]. Copyright 2014 Nature Publishing Group

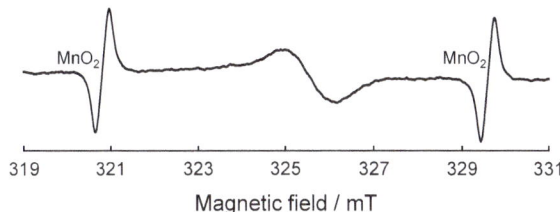

Fig. 4.3 ESR spectrum of HSA drop-casted film after irradiation of 490 MeV ^{192}Os^{30+} ion beam at a fluence of 5.0×10^9 ions cm^{-2}. Reproduced with permission from Ref. [11]. Copyright 2014 Nature Publishing Group

resonance (ESR) signal, as shown in Fig. 4.3, indicating the presence of a radical generated by the irradiation. The g-value of this radical was 2.0058, and the spectrum was similar to that of the γ-irradiated egg albumin radical [13].

The length of the nanowires generated is in direct proportion to the film thickness, and there is little variability in the length among the HSA nanowires (Fig. 4.4a, b). By adjusting the thickness of the film, nanowires of various lengths

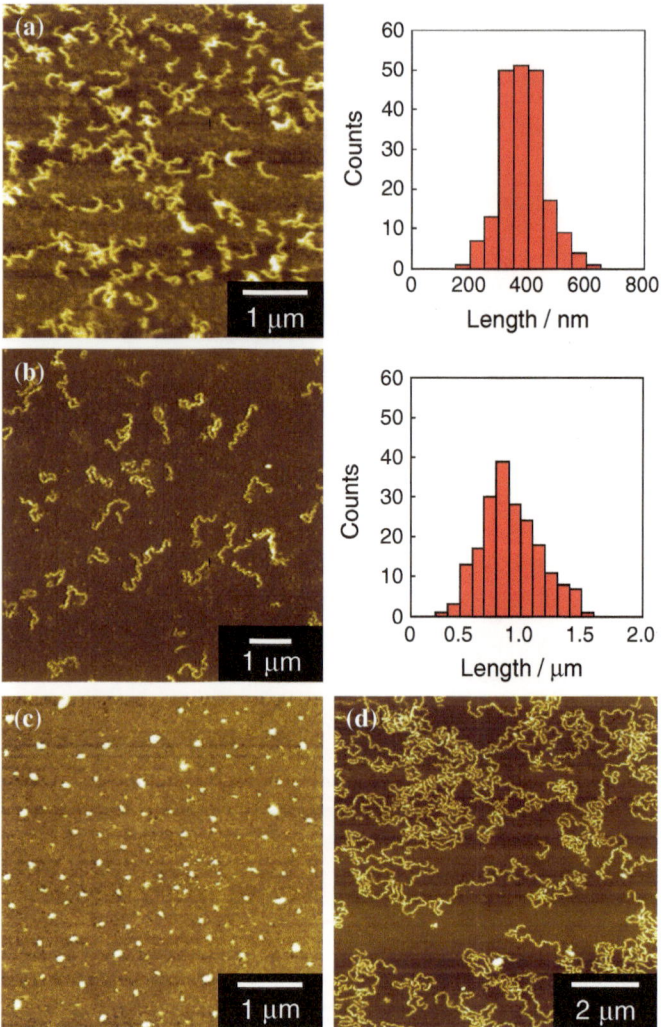

Fig. 4.4 AFM images and length distributions of the HSA nanowires formed by irradiation of **a** 500-nm-thick and **b** 1-μm-thick HSA spin-coated film with a 490 MeV $^{192}Os^{30+}$ ion beam. AFM images of **c** HSA nanodots and **d** high aspect ratio HSA nanowires (Length/Radius > 1000). Reproduced with permission from Ref. [11]. Copyright 2014 Nature Publishing Group

can be fabricated from HSA nanodots to HSA nanowires with extremely high aspect ratios of over 1,000 (Fig. 4.4c, d). In contrast, it is difficult to fabricate homogeneous protein nanowires of controlled length over a large area by the self-assembly of proteins and peptides. Thus, the SPNT method is expected to provide protein nanowires that can be used to investigate biological responses [1, 2] and have possibility to fabricate novel drug carrier and scaffolds.

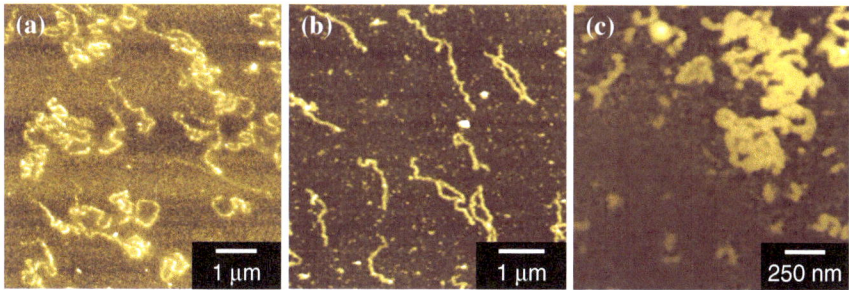

Fig. 4.5   AFM images of protein nanowires based on **a** BSA, **b** OVA, **c** avidin. Reproduced with permission from Ref. [11]. Copyright 2014 Nature Publishing Group

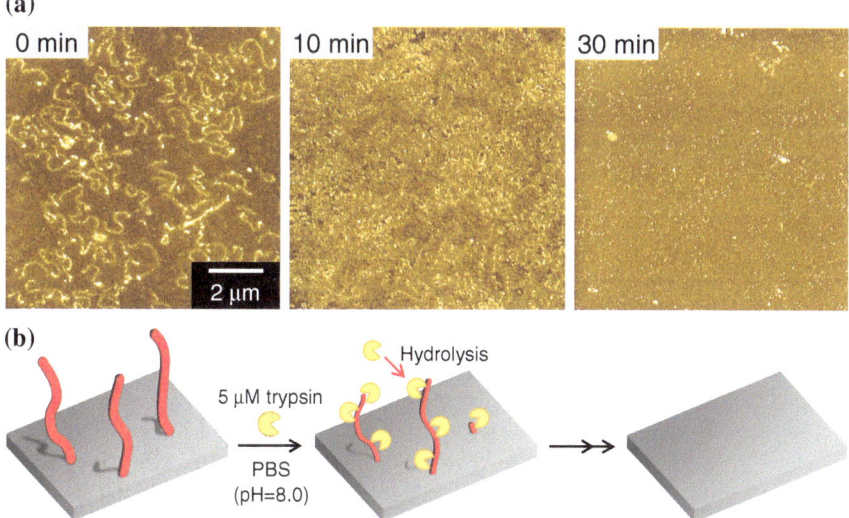

Fig. 4.6   **a** AFM images of hydrolysates of HSA nanowires after immersion in 5 μM trypsin solution at 37 °C for 0, 10, and 30 min (*left to right*). **b** Schematic depiction of the degradation of HSA nanowires by trypsin. Reproduced with permission from Ref. [11]. Copyright 2014 Nature Publishing Group

SPNT was applied to three proteins besides HSA: bovine serum albumin (BSA), ovalbumin (OVA), and avidin films. BSA and OVA films provided clearly observable BSA and OVA nanowires, similar to HSA nanowires (Fig. 4.5a, b). On the other hand, only a few nanowires were observed using avidin (Fig. 4.5c). It is likely that the avidin nanowires broke because of low mechanical strength and were removed during the development process. The formation of protein nanowires by SPNT is deeply dependent on the protein structure, namely, the amino acid sequence.

Susceptibility to degradation is a very important factor for biomaterials. Trypsin is a pancreatic serine protease that hydrolyses the peptide bond on the carboxyl side of positively charged lysine and arginine residues in peptides and proteins [14]. The HSA nanowires fabricated by SPNT are enzymatically degradable. To evaluate the enzymatic susceptibility of HSA nanowires, the HSA nanowires were incubated with trypsin and their surface was observed by AFM (Fig. 4.6a). The HSA nanowires lost their original shapes within 10 min, and no nanowires were evident following incubation for 30 min (Fig. 4.6b). This result clearly indicates that HSA nanowires are degraded by trypsin.

The structure of the ion track can be divided into two regions, i.e., the "core" and the "penumbra" [15]. The "core" is a narrow central circular zone with a radius of only a few nanometers (<1.5 nm), where extremely high-energy deposition occurs, mainly in the processes of excitation and electron plasma oscillation. The "penumbra" is a peripheral zone, with a radius ranging from nanometers to micrometers, where energy deposition occurs mainly in ionization events caused by the energetical secondary electrons released by the incident charged particle in the center of the core. In the core region, the peptide bonds and the amino acid side chains of proteins are completely destroyed, because the amount of deposited energy is too high. On the other hand, in the penumbra region, which corresponds to large part of nanowires, the amino acid residues and the structures of proteins are hardly damaged, and the HSA nanowires retain its enzymatic degradability.

4.3 Functionalization of Protein Nanowires

The avidin–biotin interaction is well known for its extraordinary affinity ($K_a = 10^{15}$ M^{-1}) [16, 17]. The surface of the HSA nanowires was modified by introducing the biotinyl group so that the nanowires could specifically interact with avidin (Fig. 4.7a). The average radii of the HSA nanowires, the biotinylated HSA nanowires, and the biotinylated HSA nanowires incubated with avidin were 8.4 ± 1.7, 9.0 ± 2.0, and 12.0 ± 2.4 nm, respectively (Fig. 4.7b–d). The increase in diameter (6.0 nm) of the biotinylated HSA nanowires corresponds to the size of avidin ($5.6 \times 5.0 \times 4.0$ nm) [18], indicating that avidin was bound to the biotinylated HSA nanowires.

The biotinylated HSA nanowires can be converted to various types of functionalized nanowires by taking advantage of the strong interaction between biotin and streptavidin and avidin. These nanowires can bind streptavidin conjugated with fluorescent probes and proteins with biological activities such as enzymes, binding proteins, and antibodies. For example, biotinylated HSA nanowires were incubated with streptavidin conjugated with Alexa Fluor 488 (fluorescent probe); after washing to remove unbound reagent, fluorescent nanowires were obtained (Fig. 4.8a). The emission spectrum of the nanowires was identical to that of the unbound fluorescent probe (Fig. 4.8b). A second example of the functionalization

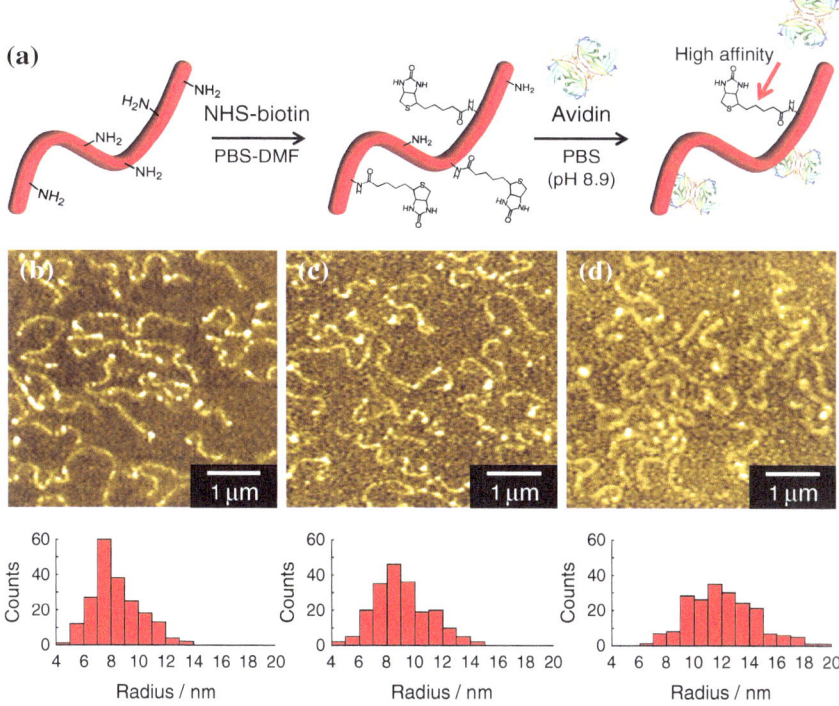

Fig. 4.7 **a** Scheme showing the surface modification of HSA nanowires by introducing biotinyl groups and the affinity between biotinylated HSA nanowires and avidin. AFM images and radius distributions of **b** HSA nanowires, **c** biotinylated HSA nanowires, and **d** biotinylated HSA nanowires after immersion in an avidin solution at room temperature for 30 min. Reproduced with permission from Ref. [11]. Copyright 2014 Nature Publishing Group

of protein nanowires was demonstrated using horseradish peroxidase (HRP), which is widely used in bioanalytical analyses due to its high substrate turnover, which provides signal amplification [19, 20]. Biotinylated HSA nanowires were treated in a manner similar to that described above with HRP conjugated with streptavidin (HRP–streptavidin) to give nanowires with peroxidase activity. HSA nanowires without biotin were also treated with HRP–streptavidin. The enzymatic activity of biotinylated HSA nanowires treated with HRP–streptavidin was 10 times larger than that of HSA nanowires treated with HRP–streptavidin. This is due to the HRP of the HRP–streptavidin complex being bound to the biotinylated nanowire through the strong interaction between the biotin moiety and the streptavidin moiety. The conjugation of proteins with biotin or streptavidin is an established technique, and many biotinylated proteins and streptavidin-conjugated proteins are commercially available. Since the biotinylated HSA nanowires can be transformed very easily into nanowires with various biological functions, these may also be of commercial interest.

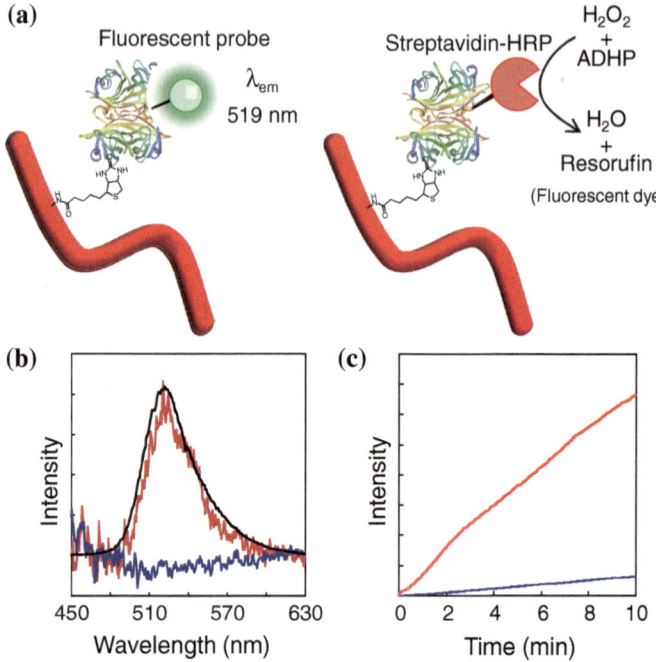

Fig. 4.8 **a** Scheme showing the surface modification of HSA nanowires by introducing bioti-nyl groups and the affinity between biotinylated HSA nanowires and avidin. AFM images and radius distributions of **b** HSA nanowires, **c** biotinylated HSA nanowires, and **d** biotinylated HSA nanowires after immersion in an avidin solution at room temperature for 30 min. Reproduced with permission from Ref. [11]. Copyright 2014 Nature Publishing Group

Fabrication of protein nanowires by irradiation is accompanied by some irra-diation damage to protein molecules. The introduced biotinyl group on the large surface of the HSA nanowires gives an ability to bind "fresh" proteins with bio-logical functions to the nanowires. If the enzymatic degradability is not required, synthetic polymeric nanowires such as poly(styrene-*co*-4-ethynylstyrene) (PSES) are available as an alternative to HSA nanowires (Fig. 4.9a). Figure 4.9b, c show the nanowires before and after modification of avidin. The radii of the nanowires were evaluated to be 16.4 and 27.1 nm, respectively, from the cross-sectional pro-files shown in Fig. 4.9d. The increase in diameter of the PESE nanowires after surface modification of avidin indicates that avidin was bound to the biotinylated PSES nanowires.

The mechanical strength of the avidin nanowires is improved by mixing avidin and HSA. Figure 4.10a shows an AFM image of HSA–avidin nanowires obtained using a 1:1 mixture of HSA and avidin. The results demonstrate that nanowires of an appropriate mechanical strength can be fabricated by mixing HSA with other proteins. To examine whether HSA–avidin nanowires exhibit specific interactions with biotin, the radius of the nanowires was measured after immersing them three

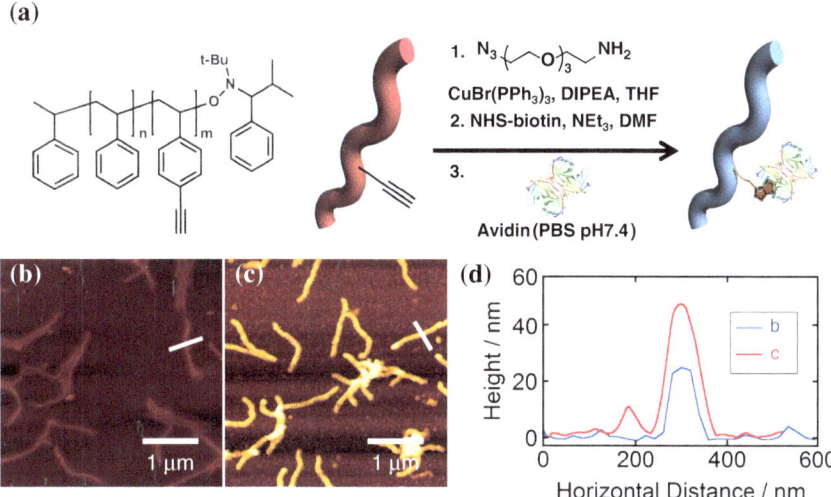

Fig. 4.9 a Chemical structure of poly(styrene-*co*-4-ethynylstyrene) (PSES) and scheme for fabrication of the protein nanowires using chemical modification of PSES nanowires and the avidin–biotin system. **b** AFM image of nanowires based on a PSES film prepared by exposure to 490 MeV $^{192}Os^{30+}$ particles at 1.0×10^8 ions cm^{-2}. **c** AFM image of modified nanowires. The surfaces were modified with avidin. **d** Cross-sectional profiles of nanowires based on PSES before and after modification. Reproduced with permission from Ref. [12]. Copyright 2012 American Chemical Society

times into dibiotinyl linker and avidin solutions. The average radius of HSA–avidin nanowires increased from 15.2 ± 2.4 to 18.3 ± 3.5 nm following this treatment (Fig. 4.10b), clearly indicating that HSA–avidin nanowires exhibit specific interactions with biotin (Fig. 4.10c). Avidin was thus immobilized through chemical linkages formed by ion beam irradiation, but its biotin-binding site of the avidin moieties was not destroyed and retained the ability to bind specifically with biotin, despite irradiation with MeV-order high-energy charged particles.

Biotinylated HSA nanowires, biotinylated synthetic polymeric nanowires, and HSA–avidin nanowires will contribute to tremendous improvement in the sensitivity of enzyme-linked immunoassay because of their large surface areas. Furthermore, they can be used for highly sensitive clinical diagnosis of diseases that include quantitative enzymes immobilization, such as inflammation system and blood coagulation system.

4.4 Fabrication of 2D Protein Sheets Based on Nanowires

Not only protein nanowires but also 2D protein sheets can be fabricated by assembly of PSES nanowires and simple surface modification. The aggregation structures of the PSES nanowires were changed with development solvents. When the

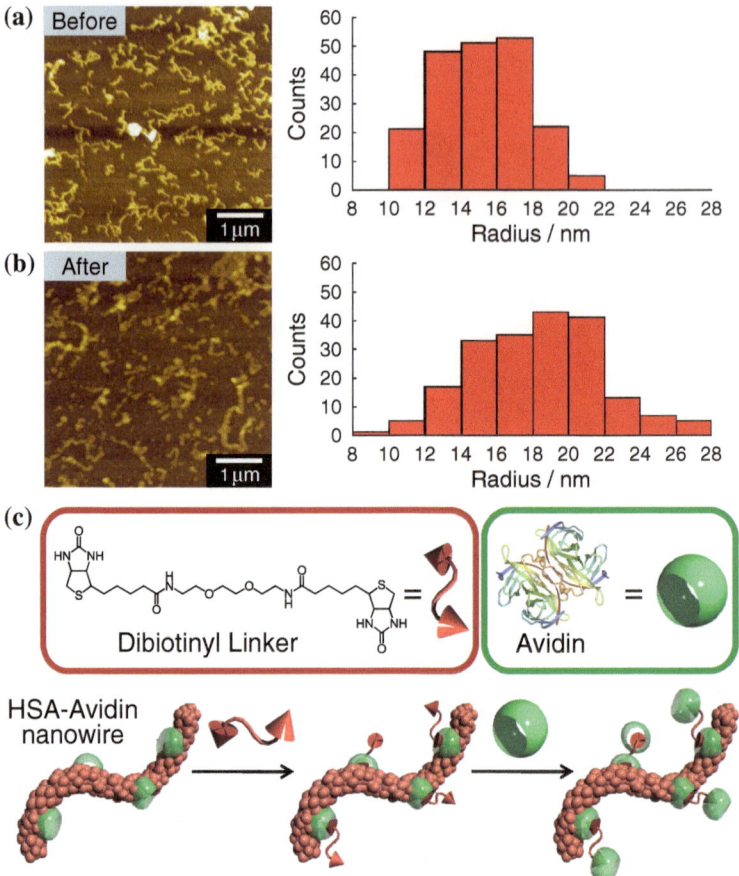

Fig. 4.10 AFM images and radius distributions of HSA–avidin (HSA: avidin = 1:1) **a** before and **b** after immersion into a dibiotinyl linker solution, followed by immersion into an avidin solution. This process was repeated three times. **c** Scheme showing the surface modification of HSA–avidin nanowires using a dibiotinyl linker and avidin. Reproduced with permission from Ref. [11]. Copyright 2014 Nature Publishing Group

development was carried out in THF and toluene, "linear" and "sheet" structures (Fig. 4.11a, b) of nanowires were confirmed, respectively. Furthermore, when the development was carried out in a mixed solvent of THF and toluene, a "net" structure (Fig. 4.11c) of the nanowires was confirmed. The arrangement of nanowires is controlled by competition of the interactions between the nanowires and substrate in each solvent. The combination of the arrangement and modification techniques makes it feasible to design and fabricate highly controlled composites on a substrate (Fig. 4.12a). After the arrangement of the nanowires, the assembly structure is fixed on the substrate (Fig. 4.12b). Avidin molecules are arranged in the order of array of nanowires (Fig. 4.12c). These 2D protein sheets might be valuable for biological assays.

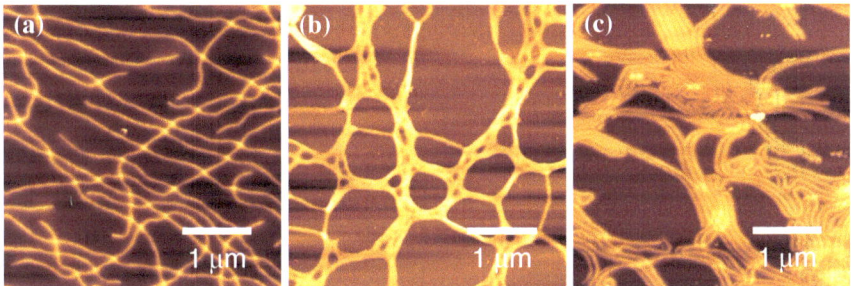

Fig. 4.11   AFM images of 8 μm PSES nanowires on a Si substrate fabricated by irradiation of 490 MeV $^{192}Os^{30+}$ particles at 1.0×10^8 ions cm^{-2}. Development was carried out in **a** THF, **b** toluene, and **c** a mixture of THF and toluene for 5 min. Reproduced with permission from Ref. [12]. Copyright 2012 American Chemical Society

Fig. 4.12   **a** Scheme for fabrication of the 2D protein sheets by modifying the assembly structure of PSES nanowires. **b** AFM image of arrays of nanowires prepared by exposing PSES film to 490 MeV $^{192}Os^{30+}$ particles at 1.0×10^8 ions cm^{-2}. Development was carried out in toluene for 5 min. **c** AFM image of arrays of protein-modified nanowires. The surface was modified with avidin. Reproduced with permission from Ref. [12]. Copyright 2012 American Chemical Society

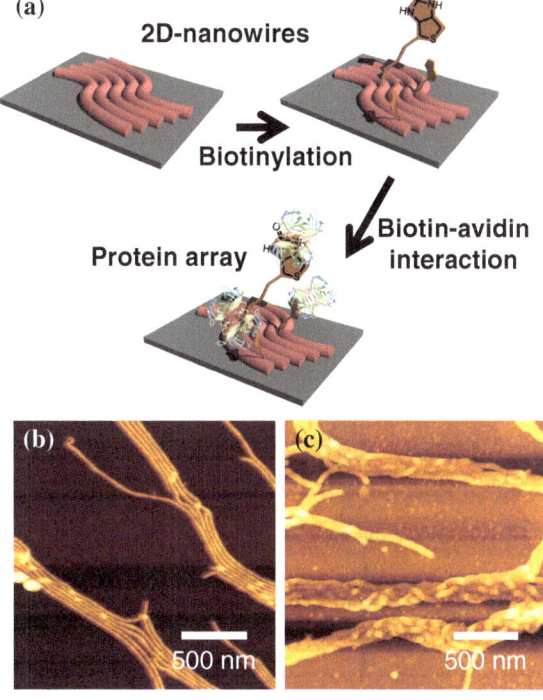

Furthermore, nanowires composed of protein and synthetic polymer were successfully fabricated by SPNT. SPNT was applied to a bilayer of HSA (upper layer) and poly(4-chlorostyrene) (P4CS) (lower layer) (Fig. 4.13a). The nanowires consist of two components, a line and a dot (Fig. 4.13b). The size and morphology of the wire are consistent with that observed for a HSA nanowire, and the dot appears

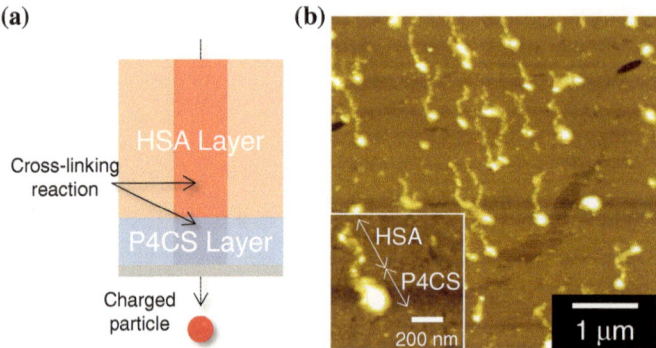

Fig. 4.13 a Schematic image of the formation of HSA–P4CS connected nanowires by SPNT from a HSA (*upper*) and P4CS (*lower*) bilayer film. **b** AFM image of the HSA–P4CS connected nanowires fabricated by exposure to a 490 MeV $^{192}Os^{30+}$ ion beam. Reproduced with permission from Ref. [11]. Copyright 2014 Nature Publishing Group

identical to an individual P4CS nanodot. It has been reported the generation of various functional nanowires such as stimuli-responsive nanowires and conductive nanowires using SPNT. Accordingly, SPNT is expected to provide homogeneous multifunctional nanowires which are otherwise difficult to fabricate, by combining a functional polymer with protein. The assemblies of protein nanowires, stimuli-responsive nanowires, and so on should contribute to the development of complex and multifunctional drug carrier and sensor.

References

1. Y.B. Lim, E. Lee, M. Lee, Angew. Chem. Int. Ed. **46**, 3475 (2007)
2. S.L. Gras, A.K. Tickler, A.M. Squires, G.L. Devlin, M.A. Horton, C.M. Dobson, C.E. MacPhee, Biomaterials **29**, 1553 (2008)
3. D. Men, Y.C. Guo, Z.P. Zhang, H.P. Wei, Y.F. Zhou, Z.Q. Cui, X.S. Liang, K. Li, Y. Leng, X.Y. You, X.E. Zhang, Nano Lett. **9**, 2246 (2009)
4. Y. Leng, H.P. Wei, Z.P. Zhang, Y.F. Zhou, J.Y. Deng, Z.Q. Cui, D. Men, X.Y. You, Z.N. Yu, M. Luo, X.E. Zhang, Angew. Chem. Int. Ed. **49**, 7243 (2010)
5. T.P. Knowles, M.J. Buehler, Nat. Nanotechnol. **6**, 469 (2011)
6. C. Foged, B. Brodin, S. Frokjaer, A. Sundblad, Int. J. Pharm. **298**, 315 (2005)
7. V. Kanchan, A.K. Panda, Biomaterials **28**, 5344 (2007)
8. R.O. Hynes, A.T. Destree, Cell **15**, 875 (1978)
9. Y. Mao, J.E. Schwarzbauer, Matrix Biol. **24**, 389 (2005)
10. K. Kadowaki, M. Matsusaki, M. Akashi, Chem. Lett. **41**, 523 (2012)
11. M. Omichi, A. Asano, S. Tsukuda, K. Takano, M. Sugimoto, A. Saeki, D. Sakamaki, A. Onoda, T. Hayashi, S. Seki, Nat. Commun. **5**, 3718 (2014)
12. A. Asano, M. Omichi, S. Tsukuda, K. Takano, M. Sugimoto, A. Saeki, S. Seki, J. Phys. Chem. C **116**, 17274 (2012)

13. J. Kumagai, K. Masui, Y. Itagaki, M. Shiotani, S. Kodama, M. Watanabe, T. Miyazaki, Radiat. Res. **160**, 95 (2003)
14. J.V. Olsen, S.E. Ong, M. Mann, Mol. Cell. Proteomics **3**, 608 (2004)
15. A. Chatterjee, H.J. Schaefer, Radiat. Environ. Biophys. **13**, 215 (1976)
16. N.M. Green, Adv. Protein Chem. **29**, 85 (1975)
17. N.M. Green, Methods Enzymol. **184**, 51 (1990)
18. L. Pugliese, A. Coda, M. Malcovati, M. Bolognesi, J. Mol. Biol. **231**, 698 (1993)
19. N.C. Veitch, Phytochemistry **65**, 249 (2004)
20. H.H. Gorris, D.R. Walt, J. Am. Chem. Soc. **131**, 6277 (2009)

Chapter 5
Stimuli-Responsive Nanomaterials

5.1 Preliminary Remarks

High aspect nano- or microstructures exhibit properties such as superhydrophobicity [1], reversible adhesion [2], structural coloration [3], antireflection [4], sensory activity [5], and actuator [6]. Despite many attempts to fabricate controlled and homogeneous structures over a large area [7–14], there have been few reports regarding nanostructures with high aspect ratios and large areas, because nanostructures are generally fragile, and consequently, the methods for fabrication are limited. Flexibility to tolerate the transformation is, particularly, needed for stimuli-responsive nanomaterials of high aspect ratios [10].

Seki and coworkers have extended the SPNT to stimuli-responsive polymer, such as poly(*N*-isopropylacrylamide) (PNIPAAm) [15] and azobenzene copolymer [16], for stimuli-responsive nanowires. PNIPAAm is a thermoresponsive polymer with a lower critical solution temperature (LCST) of approximately 32 °C in water [17, 18]. At room temperature, PNIPAAm absorbs water, which results in swelling of the polymer and transformation to a hydrophilic state, whereas at temperatures higher than the LCST, PNIPAAm shrinks and transforms to a hydrophobic state. Azobenzene is a well-known photoresponsive molecule that undergoes reversible trans-cis isomerization upon photo-irradiation or thermal heating [19–27]. The molecular length of azobenzene is changed from 9 Å (*trans*) to 6 Å (*cis*) according to the isomerization, and these conformational changes are easily detected by UV/Vis absorption spectroscopy. Here, we review the fabrication and characterization of thermoresponsive and photoresponsive nanowires by SPNT.

S. Seki et al., *High-Energy Charged Particles*,
SpringerBriefs in Molecular Science, DOI 10.1007/978-4-431-55684-8_5

5.2 Fabrication of Thermoresponsive Nanowires

A 1 μm-thick PNIPAAm-MBAAm spin-coated film (PNIPAAm/MBAAm = 100/20) was irradiated with a 490 MeV $^{192}Os^{30+}$ ion beam at a fluence of 1.0×10^8 ions cm^{-2}. The addition of the N,N'-methylene-bis-acrylamide (MBAAm) cross-linker to PNIPAAm increased the cross-linking efficiency under irradiation to maintain nanostructures. The irradiated film was developed with isopropanol, which has low surface tension, because nano- and microstructures with high aspect ratios are easily destroyed by high surface tension solvents, such as water, during the development process [28, 29]. After drying, the surface of the irradiated film was observed by atomic force microscopy (AFM). The AFM image suggested that nanowires with cross-linked structures were successfully fabricated by ion beam irradiation (Fig. 5.1a). To examine the homogeneity of PNIPAAm nanowires formed from a 1 μm-thick PNIPAAm-MBAAm film, the lengths of the PNIPAAm nanowires were estimated from AFM observation, and the length distribution is shown in Fig. 5.1b. There was little variability in the length among the PNIPAAm nanowires and the average length was 2.6 μm, which was not same as the film thickness (1 μm). The discrepancy in length must be attributable to swelling of the PNIPAAm nanowires in isopropanol, where the lengths in the swollen state were maintained due to the strong interaction between the PNIPAAm nanowires and the Si wafer during the drying process. The result indicates that highly homogeneous PNIPAAm nanowires were successfully fabricated. The diameter of the nanowires was approximately 40 nm. The relationship between the thickness of the PNIPAAm-MBAAm spin-coated film and the length of the PNIPAAm nanowires formed by ion beam irradiation was examined for a (PNIPAAm/MBAAm = 100/20) film. The nanogel length increased in direct proportion to the film thickness (coefficient of variation $R^2 = 0.989$) (Fig. 5.2a),

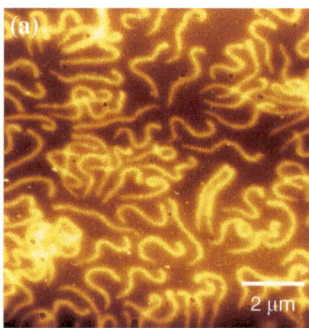
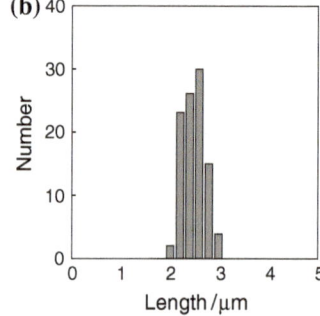

Fig. 5.1 **a** AFM image of the PNIPAAm nanowires formed by irradiation of a 1 μm-thick PNIPAAm-MBAAm spin-coated film (PNIPAAm/MBAAm = 100/20) with a 490 MeV $^{192}Os^{30+}$ ion beam at a fluence of 1.0×10^8 ions cm^{-2}. **b** Length distribution of the PNIPAAm nanowires (n = 100) formed from a 1 μm-thick PNIPAAm-MBAAm film. Reproduced with permission from Ref. [15]. Copyright 2012 American Chemical Society

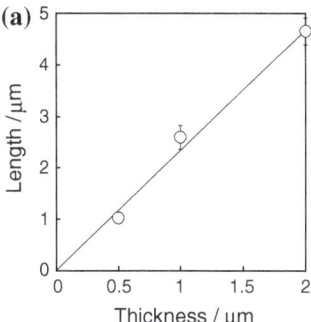
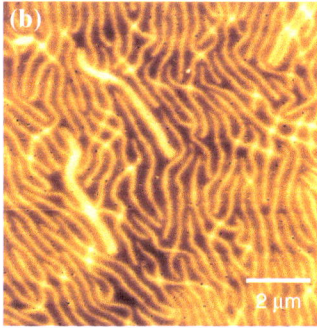

Fig. 5.2 **a** Length of the nanowires ($n = 30$) as a function of the PNIPAAm-MBAAm spin-coated film thickness. **b** AFM image of high aspect ratio nanowires (length:radius = 130:1). Reproduced with permission from Ref. [15]. Copyright 2012 American Chemical Society

which confirms that homogeneous cross-linking occurred in the film, even if the film thickness increased. For a 2 μm-thick film, the aspect ratio of the PNIPAAm nanowires formed was reached up to 130 (Fig. 5.2b).

The cross-linking efficiency of the nanowires can be easily controlled by adjusting the MBAAm content. The effects of the MBAAm content on the length and the radius of the PNIPAAm nanowires were examined using AFM (Figs. 5.3a–c). The length of the nanowires decreased with increasing MBAAm content, while the radius increased with the MBAAm content (Fig. 5.3d). It is reasonable that the length and the radius changed inversely. For a lower MBAAm content, the length was long and showed rather wide variability. It is likely that these nanowires broke in several parts because of low mechanical strength caused by low degree of cross-linking and were removed during the development process. For a higher MBAAm content, the length was short and the variability was narrow due to higher mechanical strength as a result of a higher degree of cross-linking. In general, the degree of cross-linking increased with increasing the MBAAm content of the PNIPAAm gels [30]. These results indicate that the mechanical strength of the nanowires can be controlled by adjusting the MBAAm content. Assuming spatial homogeneous distribution of MBAAm cross-linker, the estimate of diameter of nanowires derived from Eq. 3.4 depends only on the value of LET of the incident particle. The value of LET is calculated as 12,000 eV nm^{-1} for 490 MeV ^{192}Os^{30+} in a (PNIPAAm/MBAAm = 100/20) film, and the value is almost constant (<1 % decrease) over the trajectories of incident particle as long as 4 μm. This is the case giving homogeneous cross-linking in the PNIPAAm nanowires and linear correlation between aspect ratio and the initial film thickness. The cross-linking efficiency of $G(x)$ is also derived from Eq. (3.4) from observed values of r, estimated as $G(x) = 0.45, 1.6, 1.7$, and 2.8 (100 eV)$^{-1}$ for the film with varying contents of MBAAm (PNIPAAm/MBAAm) at 100/5, 100/10, 100/20, and 100/30, respectively.

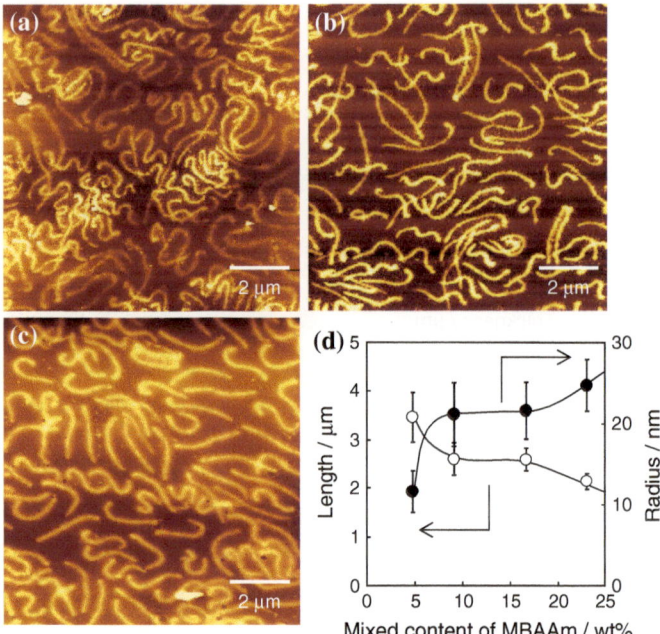

Fig. 5.3 AFM images of the PNIPAAm nanowires formed by 490 MeV $^{192}Os^{30+}$ ion beam irradiation (1.0×10^8 ions cm^{-2} fluence) of 1 μm-thick PNIPAAm-MBAAm spin-coated films with MBAAm contents (PNIPAAm/MBAAm) of **a** 100/5, **b** 100/10, and **c** 100/30. **d** Dependence of the length and radius of the nanowires on the MBAAm content of the film. Reproduced with permission from Ref. [15]. Copyright 2012 American Chemical Society

The value of $G(x)$ has been reported to depend strongly on the LET values of the incident particle, because radical coupling and/or the other second-order reactions often control the cross-linking reaction in polymer materials without cross-linkers [31, 32]. In the present case, primary the cross-links were introduced by pseudo-first-order reactions via MBAAm cross-linker molecules, giving the less dependence on the LET values.

In spite of identical set of a polymer and a cross-linker, the efficiency dramatically increases with an increase in mixed contents of MBAAm, leading to the higher density of cross-links in the PNIPAAm nanowires and mechanical strength. The nanogel with a low cross-linking density and a high aspect ratio has poor mechanical strength; however, by adjusting the film thickness and the MBAAm content, PNIPAAm nanowires with any aspect ratio and adequate mechanical strength can be fabricated.

The PNIPAAm nanowires were successfully transformed from the non-aggregated to aggregated forms over a LCST of approximately 32 °C in water. The form of the nanowires incubated in water at various temperatures was

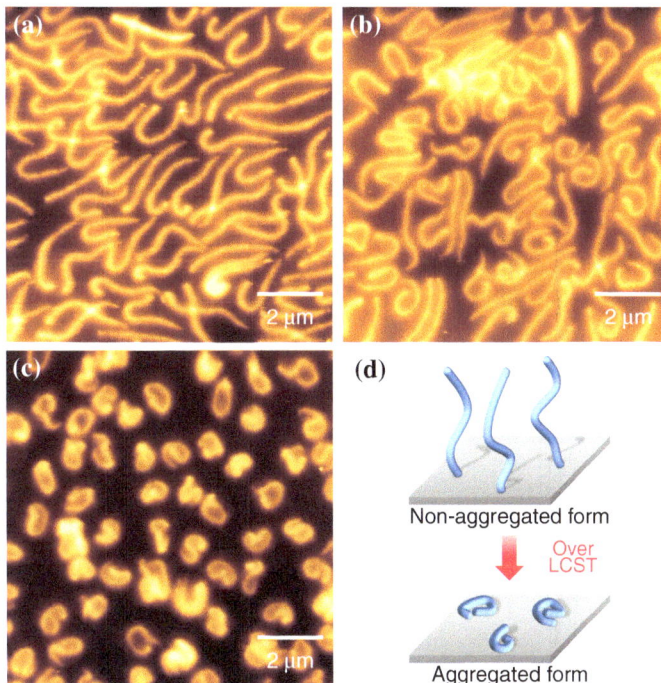

Fig. 5.4 AFM images of PNIPAAm nanowires treated in water at **a** 25 °C, **b** 40 °C, and **c** 50 °C. The nanowires were formed by irradiation of a 1 μm-thick (PNIPAAm/MBAAm = 100/20) spin-coated film with a 490 MeV ^{192}Os^{30+} ion beam (1.0×10^8 ions cm^{-2} fluence). **d** Schematic image showing the transformation of PNIPAAm nanowires from non-aggregated to aggregated forms in response to increasing temperature. Reproduced with permission from Ref. [15]. Copyright 2012 American Chemical Society

examined (Figs. 5.4a–c). The nanowires transformed from the non-aggregated form to the aggregated form with increasing temperature (Figs. 5.4d). The transformation was likely to be due to dehydration of the nanowires. The morphologies slightly changed over 40 °C (Figs. 5.4b, c), suggesting that the dehydration of the PNIPAAm nanowires is still in progress over LCST. Panels (a) and (b) in Fig. 5.5 show the length and radius distributions of nanowires incubated in water at 25, 40, and 50 °C. The average lengths at 25, 40, and 50 °C were 5.72 ± 0.40, 3.56 ± 0.65, and 3.25 ± 0.35 μm, respectively. Swollen and shrunken states were observed at 25 and 50 °C, respectively, and at 40 °C, both states were observed. The variation in length for the swollen state at 25 °C (0.40 μm) and the shrunken state at 50 °C (0.35 μm) could be attributed to the variation in the gyradius with the LCST [17, 18]. The results are compatible with the reports regarding the coil-to-globule transition of PNIPAAm. Application of the SPNT to stimulus-responsive polymers is very effective for the fabrication of nano-sized polymer actuators.

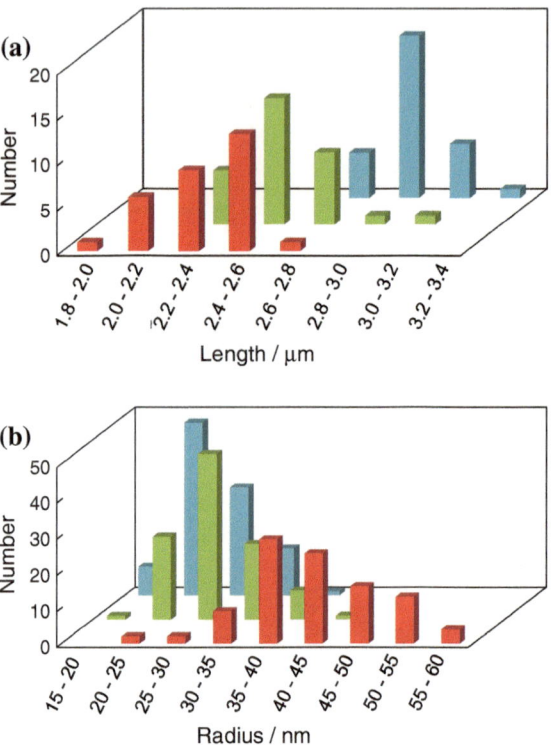

Fig. 5.5 **a** Length ($n = 30$) and **b** radius distributions ($n = 100$) of the PNIPAAm nanowires at 25 °C (*blue*), 40 °C (*green*), and 50 °C (*red*). Reproduced with permission from Ref. [15]. Copyright 2012 American Chemical Society

5.3   Fabrication of Photoresponsive Nanowires

Not only temperature-responsive nanowire but also photoresponsive nanowires can be fabricated by SPNT. As shown in Fig. 5.6, poly[(9,9′-di-*n*-octylfluorenyl-2,7-diyl)-*co*-4,4′-azobenzene] (PFOAzo) polymers comprising photochromic azobenzene and π-conjugated fluorene units were synthesized by the Suzuki-Miyaura cross-coupling reaction between 4,4′-dibromoazobenzene, 9,9′-dioctyl-2,7-dibromofluorene, and 9,9′-dioctylfluorene-2,7-diboronic acid bis(1,3-propanediol) ester. The red (PFOAzo(1:1)), orange (PFOAzo(3:1)), and yellow (PFOAzo(7:1))-colored polymers were obtained. The actual FO/Azo ratios of 1.0:1.0, 2.7:1.0, and 6.1:1.0 for PFOAzo(1:1), (3:1), and (7:1) copolymers were measured by elemental analysis, respectively. The UV-vis absorption spectra of the PFOAzo polymer films showed the slight decrease of the absorption maxima derived from the *trans*-to-*cis* isomerization of the Azo units upon UV light exposure at 401 nm (Fig. 5.7) [33, 34]. The absorption maxima of the fluorene-rich PFOAzo slightly shifted hypsochromically [416 nm for PFOAzo(1:1), 396 nm for

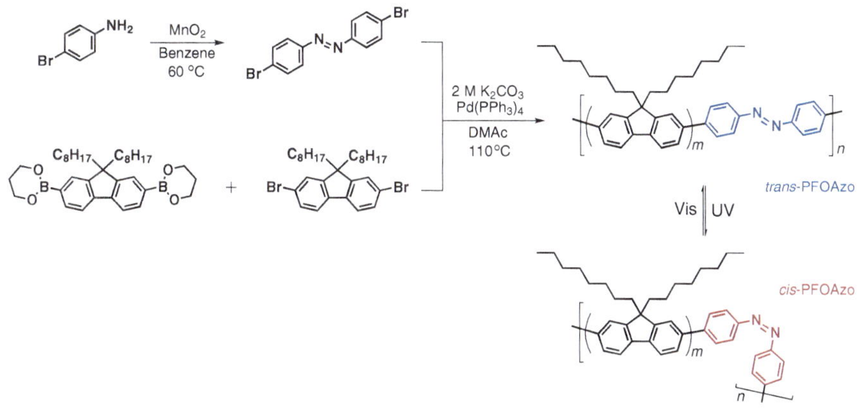

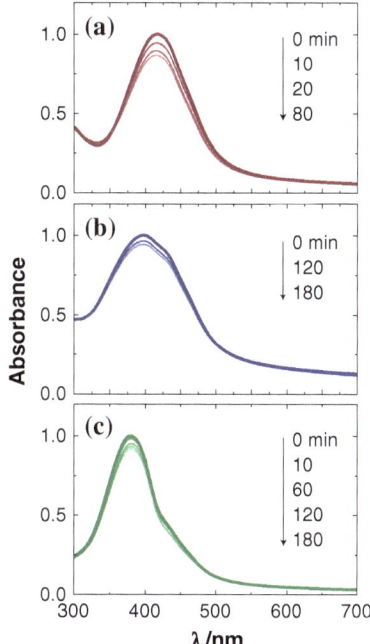

Fig. 5.6 Synthetic route to the PFOAzo polymers. Reproduced with permission from Ref. [16]. Copyright 2015 John Wiley & Sons, Inc

Fig. 5.7 Optical absorption spectra of thin films of PFOAzo **a** 1:1, **b** 3:1, and **c** 7:1 on a quartz substrate upon exposure to 401 nm light. Reproduced with permission from Ref. [16]. Copyright 2015 John Wiley & Sons, Inc

PFOAzo(3:1), and 378 nm for PFOAzo(7:1)], and the decreases in their absorptions upon *trans*-to-*cis* isomerization become less obvious gradually.

Nanowire fabrication was performed using 1 μm-thick PFOAzo(1:1, 3:1, and 7:1) films. The films were exposed to a 490 MeV ^{192}Os^{30+} ion beam at a fluence of 1.0×10^9 ions cm^{-2}, and then the nanowires were developed in an appropriate solvent by removing the unexposed areas. Through the exhaustive optimization of

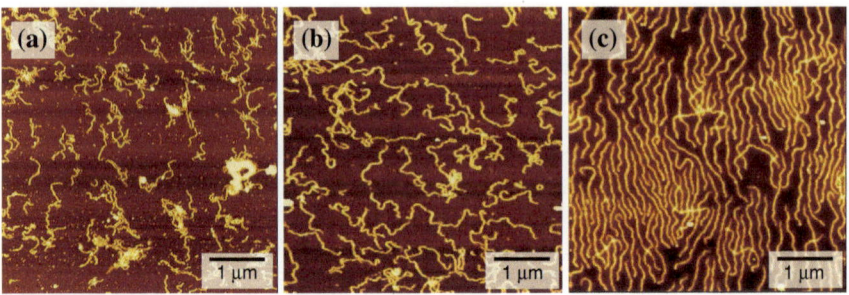

Fig. 5.8 AFM images of **a** 1:1, **b** 3:1, and **c** 7:1 PFOAzo nanowires. Reproduced with permission from Ref. [16]. Copyright 2015 John Wiley & Sons, Inc

the developer (chlorobenzene for the 1:1 composition and toluene for the others), dipping time (1–5 min), and the use of a mechanical vibrational assist, the nanowires were successfully fabricated on Si wafers for all PFOAzo(1:1, 3:1, and 7:1) (Fig. 5.8). PFOAzo(1:1) nanowires were likely to aggregate after development because of the remaining unreacted polymers (Fig. 5.8a). This is due to the insufficient solubility of PFOAzo(1:1) in chlorobenzene. Contrastingly, PFOAzo(3:1) and PFOAzo(7:1) (Figs. 5.8b, c) exhibit distinct and separate nanowires with less aggregation compared to PFOAzo(1:1), reflecting their increased solubilities. It is interesting to note that alkylfluorene-rich PFOAzo(7:1) nanowires adopted a roughly aligned morphology, probably corresponding to the direction in which the developer was pulled away. The alignment of nanowires is sometimes observed for a highly cross-linked polymer in a good solvent, and is rationalized by the high flexibility of nanowires that could be sensitive to the flow of liquid [35]. The high flexibility of nanowires is very important for reversible transformation.

Azo units in the PFOAzo(7:1) nanowires were shown reversible *trans-cis-trans* isomerization upon exposure to ultraviolet or visible light. The AFM images of the PFOAzo(7:1) nanowires just after development in toluene are shown in Fig. 5.9a. The nanowires on the Si wafer were again soaked in toluene and exposed to 401 nm light for 7 h to promote *trans*-to-*cis* isomerization. As a result, the radial of the PFOAzo(7:1) nanowires shrunk from 8.1 to 6.3 nm (Fig. 5.9b and Table 5.1). More interestingly, the shape of the nanowires changed from straight to wavy wires (the magnified images are shown in the insets of Figs. 5.9a, b). Subsequently, the *cis*-to-*trans* reversion was tested upon exposure to 499 nm light for 20 h in toluene (Fig. 5.9c). The nanowires appeared less wavy, and the radii increased to 7.3 nm. The cross section profiles of the nanowires before and after exposure to 401 and 499 nm light are shown in Fig. 5.9d. The profile data agree well with the change of the radii of the nanowires observed by AFM that resulted from the *trans*-to-*cis* and *cis*-to-*trans* isomerizations. Meanwhile, the radii of the control samples immersed in toluene for 7 h (Fig. 5.9e) and 27 h (Fig. 5.9f) without light exposure were 8.1 and 7.8 nm, respectively, which almost unchanged from that of newly developed nanowires. Therefore, the shrinkage of the radii and the change of the shape from straight to wavy form were confirmed to be caused

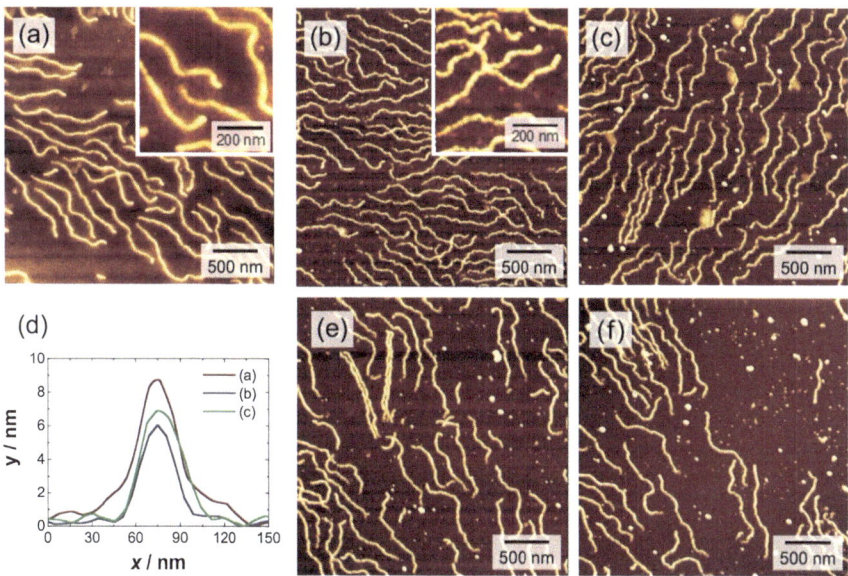

Fig. 5.9 AFM images of PFOAzo (7:1) nanowires for the sequential *trans*-to-*cis* isomerization. **a** Just after development, **b** subsequent to exposure to 401 nm light for 7 h in toluene, and **c** subsequent to 499 nm light exposure for 20 h in toluene. The *insets* of (**a**) and (**b**) are magnified images. **d** The cross section profiles of the nanowires in (**a**–**c**). Images (**e**) and (**f**) represent control experiments without light exposure: **e** dark conditions for 7 h in toluene, and **f** dark conditions for 27 h in toluene. Reproduced with permission from Ref. [16]. Copyright 2015 John Wiley & Sons, Inc

Table 5.1 Changes in PFOAzo nanowire average radii after *trans*-to-*cis* or *cis*-to-*trans* isomerization

PFOAzo	After development (nm)	*trans*-to-*cis* (control)[a] (nm)	*cis*-to-*trans* (control)[b] (nm)
7:1	8.1	6.3 (8.1)	7.3 (7.8)
3:1	7.6	5.8 (7.6)	5.5 (7.2)

The typical deviation was 10 %

[a]After exposure at 401 nm for 7 h in toluene. The control involved immersion in toluene for 7 h in the dark

[b]After subsequent exposure at 499 nm for 20 h in toluene. The control involved immersion in toluene for 20 h in the dark. Reproduced with permission from Ref. [16]. Copyright 2015 John Wiley & Sons, Inc

by isomerization rather than through a swelling effect of the solvent. On the other hand, the PFOAzo(3:1) nanowires displayed great differences in their radii (from 7.6 to 5.8 nm) upon trans-to-cis isomerization. The *cis*-to-*trans* reverse isomerization of the PFOAzo(3:1) nanowires was not successful. This is likely due to the dense packing of the *cis*-Azo units in the nanowires, preventing their re-expansion when *cis*-to-*trans* isomerization takes place.

The change of shape between *trans* (straight) and *cis* (wavy) form nanowires of PFOAzo(7:1) were quantitatively analyzed. Figure 5.10a displays surface profiles of these nanowires in their longitudinal direction. Despite the limited length scale of analysis (<150 nm) due to the incomplete stretch of nanowires, the difference in trans and cis form was clearly observed, where the roughness was almost doubled from *trans* (standard deviation, $\sigma = 0.30$ nm) to *cis* ($\sigma = 0.78$ nm). More intuitive change is evident from the statistics of end-to-end distance of the nanowires (Fig. 5.10b). The average length of the straight *trans* nanowires was 0.95 μm, while that of wavy cis nanowires decreased to 0.88 μm, as a result of shrunken shape of nanowires. Concomitantly, the distribution of the end-to-end distance was

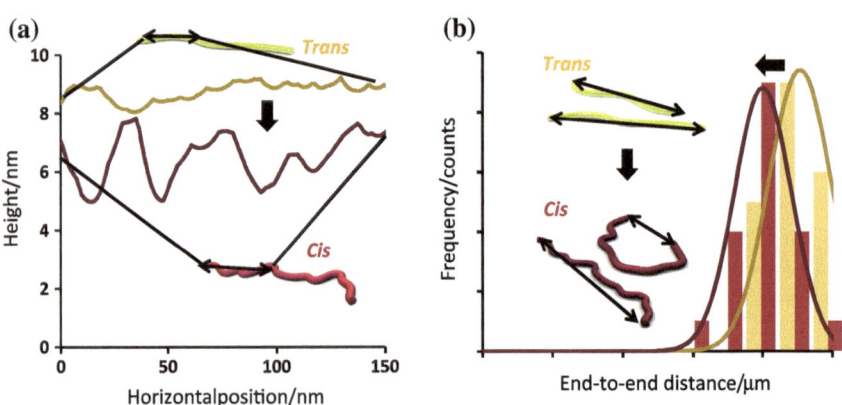

Fig. 5.10 a Surface profiles of *trans* (*upper, yellow*) and *cis* (*lower, dark red*) isomerized PFOAzo(7:1) nanowires. **b** Distribution of end-to-end distance of trans (*yellow*) and cis (*dark red*) isomerized PFOAzo(7:1) nanowires. The *solid colorized lines* are root-mean-square fits by Gaussian function. Reproduced with permission from Ref. [16]. Copyright 2015 John Wiley & Sons, Inc

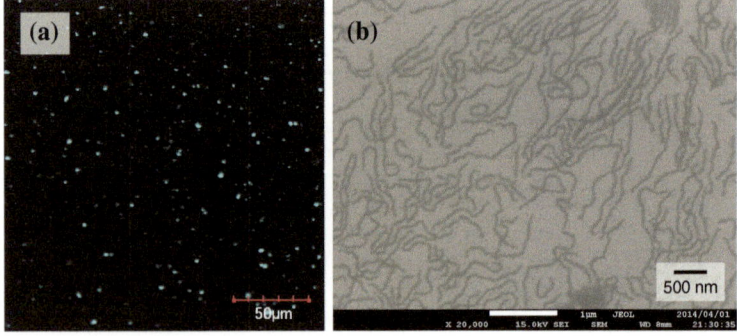

Fig. 5.11 a Fluorescence images of PFOAzo(7:1) nanowires. The fluorescence images were recorded after excitation of the sample with 405 nm light. **b** FE-SEM image of PFOAzo(7:1) nanowires. Reproduced with permission from Ref. [16]. Copyright 2015 John Wiley & Sons, Inc

widened from $\sigma = 0.07$ μm for *trans* form to $\sigma = 0.13$ μm for *cis* form. These characterizations strongly support that mechanical property of PFOAzo nanowires was controlled by exposure to ultraviolet or visible light.

The photoluminescent and semiconducting nature of PFOAzo(7:1) nanowires, presumably attributable to the π-conjugated fluorene units [36], was confirmed by fluorescence microscopy and SEM, respectively (Fig. 5.11a, b). These photoluminescence, semiconducting, and mechanical properties might offer a pathway toward rendering organic nanowires as excellent candidates for light-controlled electromechanical applications and unidirectional manipulation along the nanowires with unpolarized light.

References

1. Y. Leng, H.P. Wei, Z.P. Zhang, Y.F. Zhou, J.Y. Deng, Z.Q. Cui, D. Men, X.Y. You, Z.N. Yu, M. Luo, X.E. Zhang, Angew. Chem. Int. Ed. **49**, 7243 (2010)
2. T.P. Knowles, M.J. Buehler, Nat. Nanotechnol. **6**, 469 (2011)
3. C. Foged, B. Brodin, S. Frokjaer, A. Sundblad, Int. J. Pharm. **298**, 315 (2005)
4. V. Kanchan, A.K. Panda, Biomaterials **28**, 5344 (2007)
5. R.O. Hynes, A.T. Destree, Cell **15**, 875 (1978)
6. Y. Mao, J.E. Schwarzbauer, Matrix Biol. **24**, 389 (2005)
7. P. Lalanne, G.M. Morris, Nanotechnology **8**, 53 (1997)
8. A.K. Geim, S.V. Dubonos, I.V. Grigorieva, K.S. Novoselov, A.A. Zhukov, S.Y. Shapoval, Nat. Mater. **2**, 461 (2003)
9. Z.-Z. Gu, H. Uetsuka, K. Takahashi, R. Nakajima, H. Onishi, A. Fujishima, O. Sato, Angew. Chem. Int. Ed. **42**, 894 (2003)
10. B.A. Evans, A.R. Shields, R.L. Carroll, S. Washburn, M.R. Falvo, R. Superfine, Nano Lett. **7**, 1428 (2007)
11. Y.F. Huang, S. Chattopadhyay, Y.J. Jen, C.Y. Peng, T.A. Liu, Y.K. Hsu, C.L. Pan, H.C. Lo, C.H. Hsu, Y.H. Chang, C.S. Lee, K.H. Chen, L.C. Chen, Nat. Nanotechnol. **2**, 770 (2007)
12. J. Toonder, F. Bos, D. Broer, L. Filippini, M. Gillies, J. de Goede, T. Mol, M. Reijme, W. Talen, H. Wilderbeek, V. Khatavkar, P. Anderson, Lab Chip **8**, 533 (2008)
13. H. Yabu, Y. Hirai, M. Kojima, M. Shimomura, Chem. Mater. **21**, 1787 (2009)
14. L. Feng, Y. Zhang, M. Li, Y. Zheng, W. Shen, L. Jiang, Langmuir **26**, 14885 (2010)
15. M. Omichi, H. Marui, K. Takano, S. Tsukuda, M. Sugimoto, S. Kuwabata, S. Seki, A.C.S. Appl, Mater. Interfaces **4**, 5492 (2012)
16. H.L. Cheng, M.T. Tang, W. Tuchinda, K. Enomoto, A. Chiba, Y. Saito, T. Kamiya, M. Sugimoto, A. Saeki, T. Sakurai, M. Omichi, D. Sakamaki, S. Seki, Adv. Mater. Interfaces **2**, 1400450 (2015)
17. X. Wang, X. Qiu, C. Wu, Macromolecules **31**, 2972 (1998)
18. C. Wu, X. Wang, Phys. Rev. Lett. **80**, 4092 (1998)
19. G.S. Kumar, D.C. Neckers, Chem. Rev. **89**, 1915 (1989)
20. A. Natansohn, P. Rochon, Chem. Rev. **102**, 4139 (2002)
21. H.M.D. Bandara, S.C. Burdette, Chem. Soc. Rev. **41**, 1809 (2012)
22. D.R. Wang, X.G. Wang, Prog. Polym. Sci. **38**, 271 (2013)
23. M.M. Russew, S. Hecht, Adv. Mater. **22**, 3348 (2010)
24. H.F. Yu, T. Ikeda, Adv. Mater. **23**, 2149 (2011)
25. T. Ikeda, J. Mamiya, Y.L. Yu, Angew. Chem. Int. Ed. **46**, 506 (2007)
26. G.M. Spinks, Angew. Chem. Int. Ed. **51**, 2285 (2012)

27. S. Iamsaard, S.J. Aßhoff, B. Matt, T. Kudernac, J.L.M. CornelissenJeroen, S.P. Fletcher, N. Katsonis, Nat. Chem. **6**, 229 (2014)
28. T. Tanaka, M. Morigami, N. Atoda, Jpn. J. Appl. Phys. **32**, 6059 (1993)
29. A. del Campo, E. Arzt, Chem. Rev. **108**, 911 (2008)
30. H. Senff, W. Richtering, Colloid Polym. Sci. **278**, 830 (2000)
31. S. Seki, S. Tsukuda, K. Maeda, S. Tagawa, H. Shibata, M. Sugimoto, K. Jimbo, I. Hashitomi, A. Kohyama, Macromolecules **38**, 10164 (2005)
32. S. Seki, S. Tagawa, Polym. J. **39**, 277 (2007)
33. R. Zhao, X. Zhan, J. Yao, G. Sun, Q. Chen, Z. Xie, Y. Ma, Polym. Chem. **4**, 5382 (2013)
34. N. Anwar, T. Willms, B. Grimme, A.J.C. Kuehne, ACS Macro Lett. **2**, 766 (2013)
35. A. Asano, M. Omichi, S. Tsukuda, K. Takano, M. Sugimoto, A. Saeki, S. Seki, J. Phys, Chem. C **116**, 17274 (2012)
36. S. Seki, A. Saeki, W. Choi, Y. Maeyoshi, M. Omichi, A. Asano, K. Enomoto, C. Vijayakumar, M. Sugimoto, S. Tsukuda, S. Tanaka, J. Phys. Chem. B **116**, 12857 (2012)

Chapter 6
Nanowires for Renewable Energy

6.1 Preliminary Remarks

Through diminishing fossil fuels in the past century, we come to notice the urgent requirement of carbon-free energy, leading to a global exploration of renewable energy source to meet rising demands. Photovoltaic cell made of Si and compound semiconductors is a premiere technology for converting the sunlight energy to electricity; however, the cost per unit energy production including manufacture, maintenance, and real estate for installment and operation has not been satisfactory yet. As a counterpart of inorganic solar cell, organic photovoltaic (OPV) cells have emerged to offer a low-cost renewable source compatible with cost-effective roll-to-roll process [1–10]. Of unique advantage is a wide freedom in color and shape management, i.e., transparent OPV can be placed on a roof of green house without disturbing growth of inside plants, colorful OPV is acceptable as a stylish interior and exterior, and light-weight OPV can be transferred into mostly desirable shape such as a round roof of car and a telegraph pole. Despite these intrinsic benefits, OPV is still in a developing regime from the viewpoints of power conversion efficiency (PCE) and stability [11].

Bulk heterojunction (BHJ)-based OPV consisting of positive (p)-type and negative (n)-type materials is a notable architecture [12]. This allows the formation of a bi-continuous network with a large p/n interface, so that PCE can be improved. Fullerene, which has spherical π-conjugation and intriguing electronic nature, is the most privileged n-type material, which also provides wide applications in cosmetics, fluid lubrication, and hard plastics [13–15]. The marked progress of fullerene derivatives has been leveraged due to their versatile synthesis and high PCE in OPV performance [16–24], while the thermal instability of fullerene phase is one of the major concerns for the practical application of OPV. Therefore, high-conductive fullerene nanowires in BHJ architecture might merit both a large interface of p/n junctions and electron transport to the cathode.

© The Author(s) 2015
S. Seki et al., *High-Energy Charged Particles*,
SpringerBriefs in Molecular Science, DOI 10.1007/978-4-431-55684-8_6

Nanostructured OPVs have attracted a great deal of attention for improving efficiency and long-term stability, because nanostructured semiconductors could provide an efficient pathway for charge transport and reduce the excessive inter-mixing of materials under high-temperature operation. In fact, BHJ network is formed as a result of delicate kinetic and thermodynamic processes of p- and n-type semiconductors, with an aid of solvent and additive [25, 26]. Therefore, a simple blend of two types of semiconductors inherently comprises an issue regarding thermal stability; in particular, diffusive motion of molecular fullerenes must be retarded in commercial OPVs. Block copolymer might be a smart design to form thermally stable nanostructure, and thus researchers have developed p-bridged-n [27–32] (Fig. 6.1a, b) or surfactant block copolymer [33]. However, PCEs are low or moderate, mainly because the surface area of donor/acceptor interface is much smaller for a cylinder or lamellar structure than BHJ, resulting in a low short-circuit current density (J_{sc}). Photo and electron beam lithographies [34] as well as imprinting [35], one of the most important technologies for nanometer-scaled patterning of ultra-large-scale integration circuit, have been introduced into OPV (Fig. 6.1c). However, this top-down patterning is costly and not so effective in OPV, while bottom-up approach utilizing self-assembling nature is rather powerful, for example, formation of pillar-shaped porphyrin mediated by soluble fullerene led to a high PCE (Fig. 6.1d) [36]. Metal nanoparticles [37] and nanorods [38] are compelling materials in optical management of incident light. They can serve as an antenna for light harvesting via surface plasmon resonance, a scattering media to extend the light path, and a transparent electrode without significant loss in conductivity (Fig. 6.1e) [39]. Nanostructured dye such as phthalocyanine [40] is another notable example that allows for enhanced photoabsorption in the long wavelength region and contributes to an increase of J_{sc} (Fig. 6.1f).

As reviewed in this book, single-particle nanofabrication technique (SPNT) can be conceived to be applied to the nanostructured OPV (Fig. 6.2). Although many types of conjugated polymer nanowires have been successfully formed by SPNT [41–44], its applicability to conjugated molecules has remained elusive, because organic molecules require a large number of cross-linking reactions to ensure insolubility against the developer. However, fullerenes were found to form nanowires, owing to unprecedentedly high cross-linking efficiency. Notably, we demonstrated universal applicability of SPNT to the fullerene derivatives including C_{60}, methanofullerenes of C_{61} and C_{71}, and indene C_{60} bis-adduct [45]. A chain polymerization reaction initiated by exposure to a high-energy ion beam yielded nanowires with 8–11 nm in radii and a few tens to thousands of nanometers in length.

Herein, we review the use of [6,6]-phenyl C_{61} butyric acid methyl ester (PCBM) and indene-C_{60} bis adducts (ICBA) nanowires in a prototype poly(3-hexylthiophene) (P3HT)-based OPV. By including fullerene nanowires with precisely controlled length and density, the power conversion efficiency was successfully improved. Importantly, whether a device structure is a normal (top cathode, bottom anode) or an inverted (top anode, bottom cathode) type has a crucial impact on the OPV output. P3HT nanowires were also produced, but found to simply demerit the OPV performance, probably due to the significant polymer damage upon irradiation of high-energy ion beam.

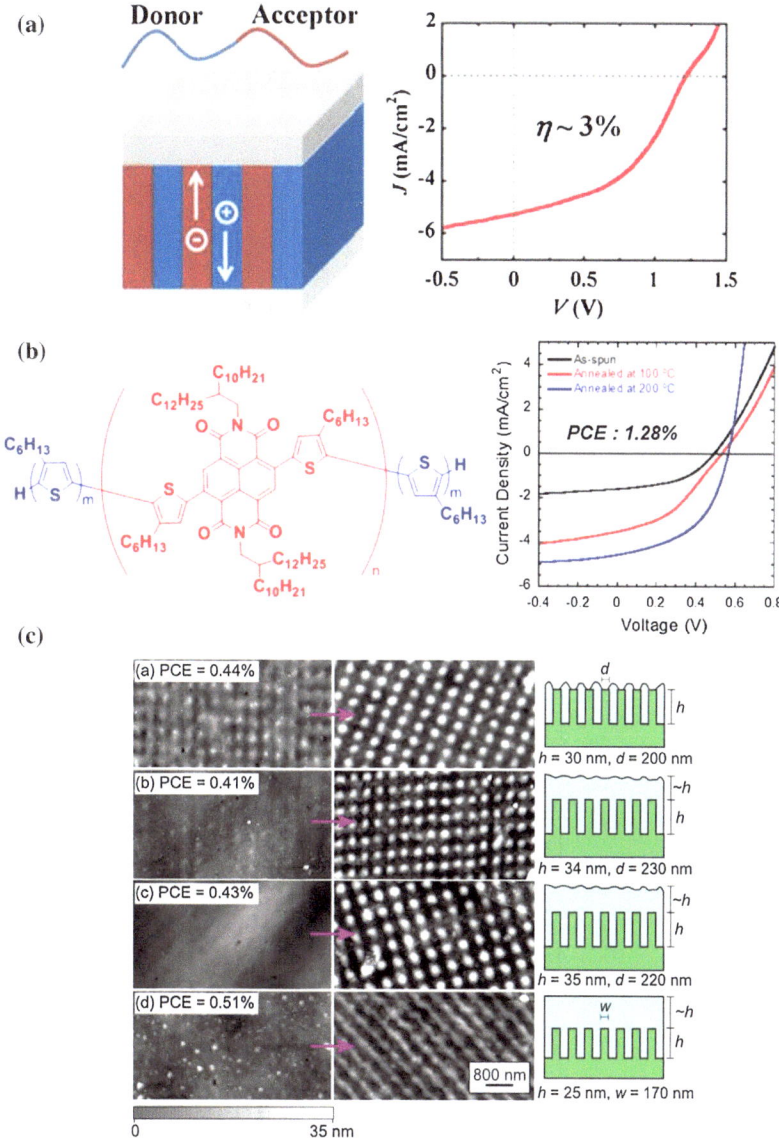

Fig. 6.1 Examples of nanostructured OPV. **a** Di-block copolymer OPV of P3HT-*b*-PFTBT. Reproduced with permission from Ref. [27]. Copyright 2013 American Chemical Society. **b** Tri-block copolymer OPV of P3HT-PNBI-P3HT. Reproduced with permission from Ref. [28]. Copyright 2012 American Chemical Society. **c** P3HT pillar fabricated by electron beam lithography and filled by PCBM. Reproduced with permission from Ref. [34]. Copyright 2012 American Chemical Society. **d** Columnar structure of tetrabenzoporphyrin and silylmethyl[60]fullerene OPV. Reproduced with permission from Ref. [36]. Copyright 2009 American Chemical Society. **e** Transparent OPV using Ag nanorod. Reproduced with permission from Ref. [39]. Copyright 2012 American Chemical Society. **f** Fluorinated copper phthalocyanine nanowires for efficient OPV. Reproduced with permission from Ref. [40]. Copyright 2012 American Chemical Society

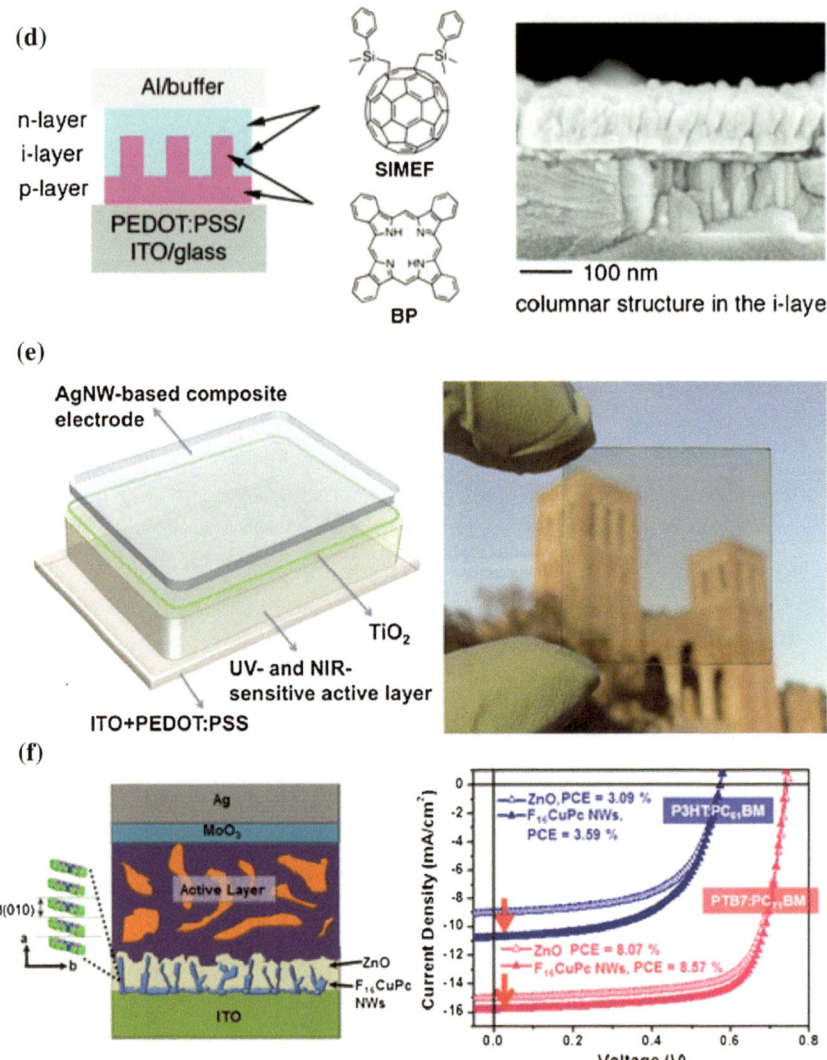

Fig. 6.1 (continued)

6.2 Universal Formation of Fullerene Nanowires by SPNT

Since the discovery of Buckminster Fullerene in 1985, a flourishing evolution has been witnessed regarding their synthesis, characterization, and industrial application [46], Among them, [6,6]-phenyl C_{61} butyric acid methyl ester ($PC_{61}BM$) [47], [6,6]-phenyl C_{71} butyric acid methyl ester ($PC_{71}BM$), and indene-C_{60} bis-adduct (ICBA) [48] illustrated in Fig. 6.3 are the most famous and commercially available

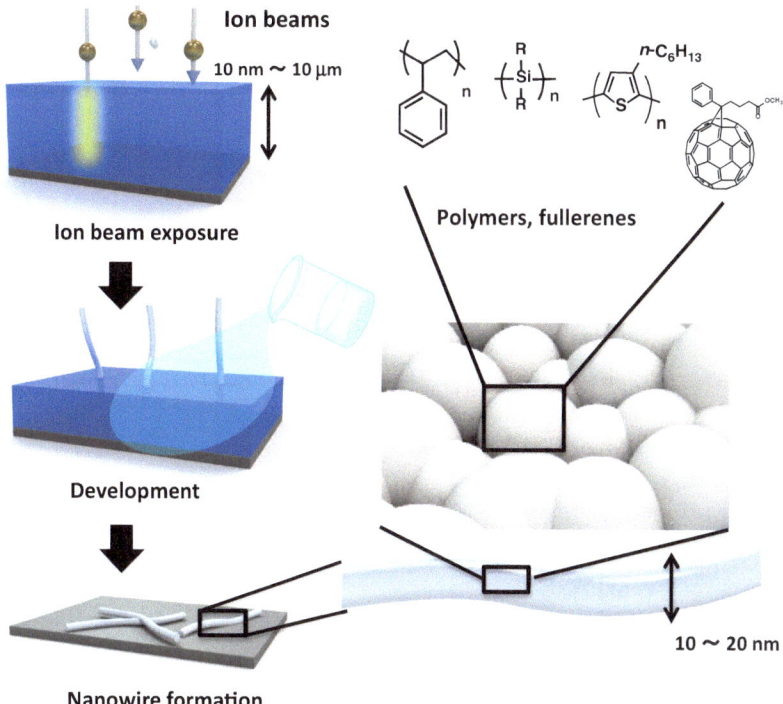

Fig. 6.2 Schematic of polymer and fullerene nanowires formation by SPNT

soluble fullerenes as an efficient *n*-type material in BHJ. Therefore, SPNT using 450 MeV Xe ions was applied to these three fullerenes as well as pristine C_{60} film, of which films were prepared by drop casting or thermal evaporation in a vacuum chamber, respectively. After developing unexposed area in toluene, $PC_{61}BM$ nanowires were successfully formed, as visualized by atomic force microscopy (AFM) images (Fig. 6.3). The density of the nanowires can be facilely controlled by just increasing the ion beam fluence. Importantly, the lengths of the nanowires are mostly equal to the thickness of the original film, irrespective of the ion fluence. Moreover, the number of nanowires coincided with the number of incident ions calculated from the corresponding fluence. In light of these observations, we can safely conclude that (1) respective ion trajectory forms one nanowire, and (2) nanowire is neither removed from the substrate nor swollen in during the development process. Therefore, a precise control over the length and density is possible by film thickness and fluence, respectively.

Similarly, all of the C_{60}, $PC_{61}BM$, $PC_{71}BM$, and ICBA fullerene derivatives were formed into thick and straight nanowires as evidenced in the AFM images (Fig. 6.3). Based on the ellipsoidal model and the measurement of one hundred points of height and width, the distribution of the nanowire radius was evaluated and shown as the histogram below each micrograph. ICBA has been extensively used

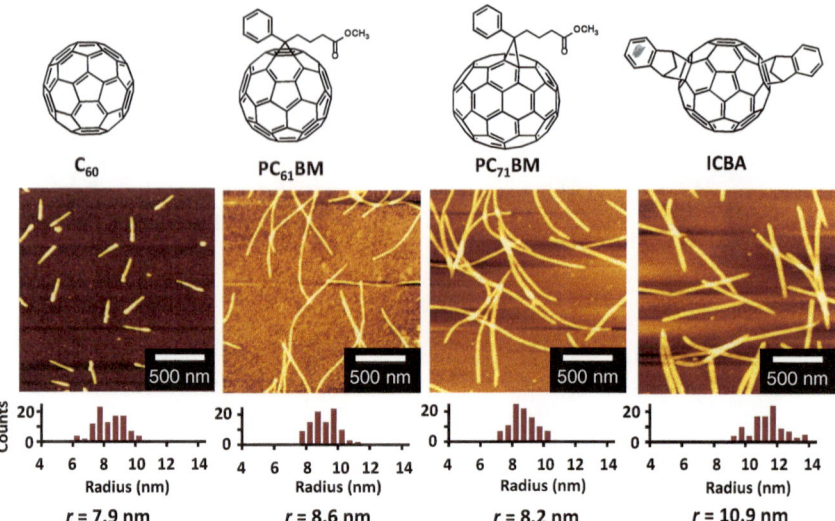

Fig. 6.3 Chemical structures of fullerene derivatives and their nanowire images measured by AFM. The *bottom* histogram shows the distribution of nanowire radius based on ellipsoidal model. The average radius, *r*, is indicated. Reproduced with permission from Ref. [45]. Copyright 2012 Nature Publishing Group

in BHJs in combination with P3HT, because the shallow lowest unoccupied molecular orbital (LUMO) achieves a high open-circuit voltage (V_{oc}) for the OPV [48]. The averaged radius of ICBA is slightly larger (11 nm) than the other fullerenes (8–9 nm), probably because the presence of ICBA regioisomers may disrupt three-dimensional crystallization [20]. Mixture of regioisomers could lead to increased radius and flexibility of the nanowires that reflect the packing in the fullerene film and its solubility. In contrast, the topography of the 2-μm-length nanowires made of bare C_{60} is obviously rigid with a complete straight shape, collaborating the presumed influence of substituent and solubility on the radius and shape.

Molecular scale surface morphology of a C_{60} nanowire was investigated by scanning tunneling microscopy (STM), where closely packed fullerene molecules are clearly visualized (Fig. 6.4a). Dimerization and polymerization reactions of C_{60} had been a hot topic in 1990s and reported with light exposure [49, 50], high temperature and pressure [51], and γ-ray irradiation [52]. From the infrared and Raman spectroscopic studies [50], the 2+2 cycloaddition reaction has been proposed for the photopolymerization reaction (Fig. 6.4b). Taking account of these reports and the good mechanical strength of the nanowires, SPNT is also expected to cause a similar polymerization reaction, in which the substituents of fullerene have no direct effect on the polymerization.

The radiation-induced gelation theory reviewed in the Chap. 3 is useful for quantitative analysis of nanowire radius (r) on the basis of a given cross-linking efficiency, $G(x)$. According to Eq. 3.4, this in turn can be used to calculate $G(x)$ from the evaluated nanowire radius. Accordingly, $G(x)$ is expressed as [53]

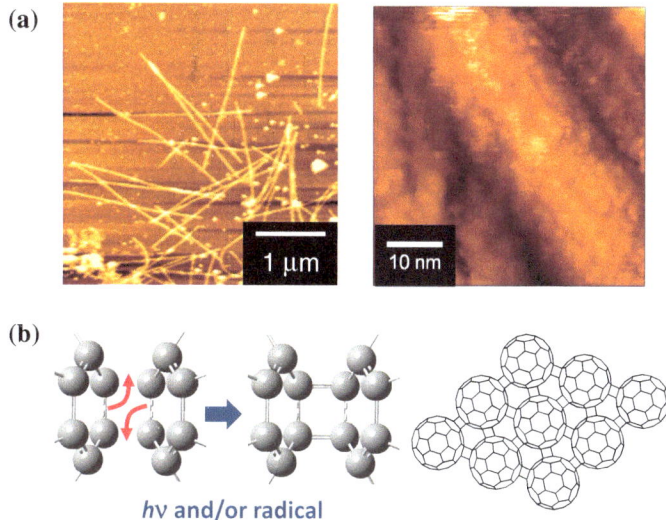

Fig. 6.4 **a** *Left* AFM image of long and straight C_{60} nanowires. *Right* STM image of C_{60} nanowires. **b** Proposed dimerization reaction via 2+2 cycloaddition. The *right* schematic shows the expected structure of the polymerized fullerenes. Reproduced with permission from Ref. [45]. Copyright 2012 Nature Publishing Group

$$G(x) = \frac{400\pi\rho N_A \cdot r^2}{\mathrm{LET} \cdot M} \ln\left(\frac{e^{1/2}r_p}{r_c}\right). \tag{6.1}$$

The validity of this equation has been proved for many types of polymers [41–43]. The typical $G(x)$s among synthetic polymers is ranging from 0.30 to 1.6 $(100\ \mathrm{eV})^{-1}$ [54]. However, $G(x)$ of the fullerene nanowires was as high as 29–55 $(100\ \mathrm{eV})^{-1}$, approximately two orders of magnitude larger than those of typical polymers. More surprisingly, these $G(x)$ values exceed the initial ionization yield (ca. 5 $(100\ \mathrm{eV})^{-1}$), suggestive of the chain polymerization reaction mediated by radical.

6.3 *p/n* Heterojunction Nanowires

Since nanowire is formed in the direction perpendicular to the substrate, one might suggest an attempt to fabricate heterojunction nanowires from a double-layer film. *p/n* heterojunction nanowire is a promising nanostructure for molecular electronics, and thus we examined double layers comprising *p*-type poly(9,9′-*n*-dioctylfluorene) (PFO) and *n*-type $PC_{61}BM$ (Fig. 6.5a) [45]. Formation of a discrete double layer without significant inter-mixing is a requisite for this purpose. We found that such a layer is formed by spin-coating toluene solution of PFO onto

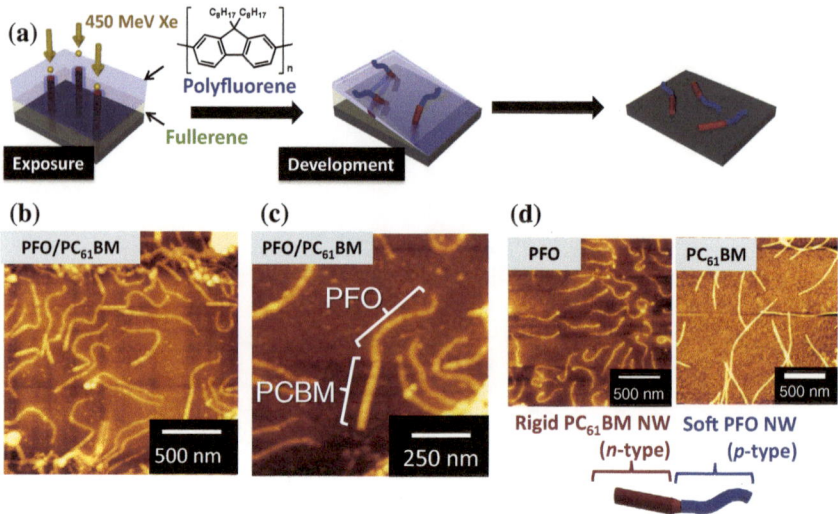

Fig. 6.5 a Schematic showing the formation of *p–n* junction nanowires by SPNT from a PFO (*upper*) and PC$_{61}$BM (*lower*) bilayer. **b, c** AFM images of PFO/PC$_{61}$BM nanowires. **d** AFM image of separately produced PFO and PC$_{61}$BM nanowires. Reproduced with permission from Ref. [45]. Copyright 2012 Nature Publishing Group

a less soluble PC$_{61}$BM layer prepared by drop casting. The nanowire formation is clearly dictated from the AFM image in Fig. 6.5b. A magnified image of the nanowire displays that the nanowire consists of a straight, thick component and a thin, winding component (Fig. 6.5c). Interestingly, the size of the former component is consistent with that observed for a PC$_{61}$BM nanowire, and the latter is identical to a PFO nanowire (Fig. 6.5d). Furthermore, the aspect ratio (width/height) of the PC$_{61}$BM nanowire cross section is as small as 2.3, while that of PFO nanowires is 7.3. This collaborates the higher rigidity of fullerene nanowires than PFO nanowires. As is often the case with polymer nanowires, the PFO nanowires are flattened by either their own weight or by adhesive interaction with the surface. We should emphasize that SPNT was proved versatile for producing any type of heterojunction between two dissimilar materials as long as a double layer is prepared. The wide applicability of SPNT is contrast to the self-assembling technology that requires subtle and advanced synthetic procedure [55].

6.4 Photovoltaic Application of Nanowires

The *p*/*n* heterojunction demonstrated might be ideal as a nanoscale diode for molecular switching; however, macroscopic integration of the *p*/*n* nanowires into the device still faces big technical problems to be solved. Therefore, the first step toward OPV application is the hybridization of PC$_{61}$BM nanowires with

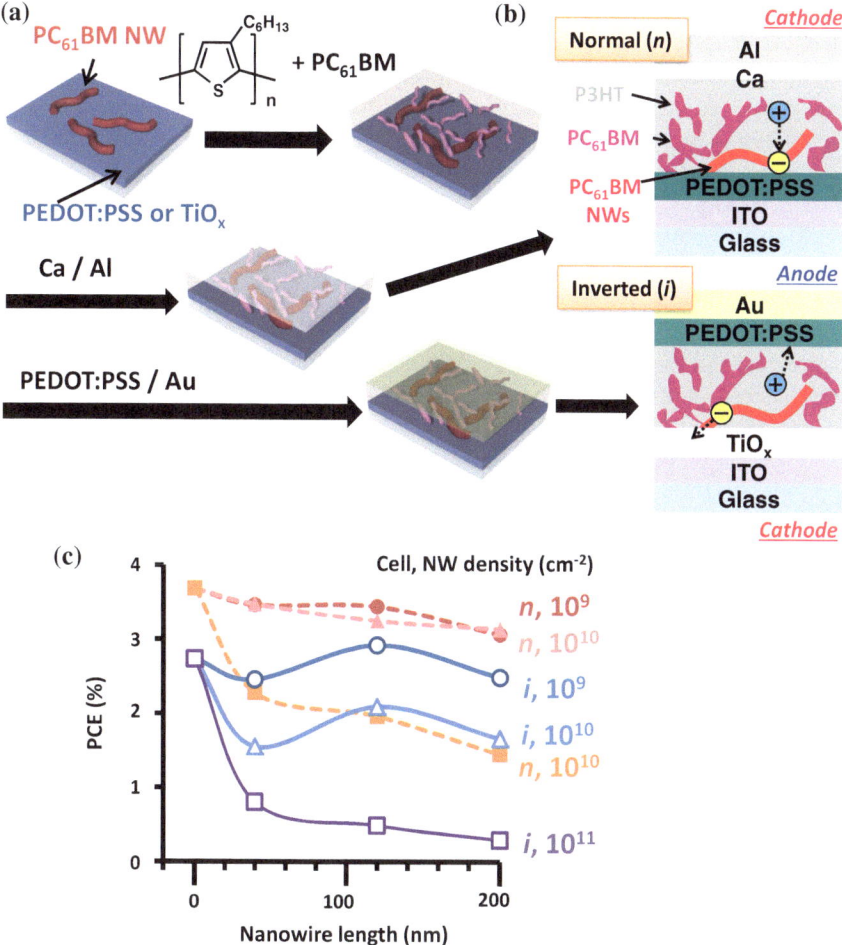

Fig. 6.6 **a** Schematic diagram of OPV fabrication. **b** Normal (*n*) and inverted (*i*) OPV structures. $PC_{61}BM$ NWs are concentrated on the lower layer. **c** *J–V* curves for the OPV devices under AM 1.5G pseudosunlight (100 mW cm^{-2}). *Dotted* and *solid lines* represent the normal and inverted structures, respectively. Reproduced with permission from Ref. [45]. Copyright 2012 Nature Publishing Group

BHJ-type OPV [45]. Figure 6.6a, b illustrates schematics for the hybridization of $PC_{61}BM$ nanowires into the benchmark BHJ of P3HT:$PC_{61}BM$. In a normal cell, poly(3,4-ethylenedioxythiophene):poly(styrenesulfonate) (PEDOT:PSS) was used as the anode buffer layer, while in an inverted cell, TiO_x was used as the cathode buffer layer. They were coated on a cleaned indium tin oxide (ITO)/ glass substrate. Subsequently, $PC_{61}BM$ nanowires, of which length and density were controlled by the $PC_{61}BM$ film thickness and ion beam fluence, respectively, were fabricated. We confirmed that $PC_{61}BM$ nanowires were not removed from the buffer layer during the development process. And then the active layer of P3HT:PC$_{61}$BM = 1:1 w/w was spin-coated from o-dichlorobenzene (oDCB) solution in a nitrogen glove box, followed by solvent annealing in a half-covered petri dish and thermal annealing at 150 °C for 10 min. A Ca/Al cathode and PEDOT:PSS/Au anode layers were subsequently deposited in a vacuum chamber or spin-coating for the normal and inverted cells, respectively. Figure 6.6c shows current density–voltage (J–V) curves for P3HT:PC$_{61}$BM devices with and without PC$_{61}$BM nanowires under AM 1.5 G pseudosunlight. In the case of normal cells, the PCE was simply decreased from 3.68 % without nanowires to 3.06 % at nanowire density (ρ_{NW}) = 10^9 cm^{-2} and to 1.44 % at ρ_{NW} = 10^{11} cm^{-2}. The impairing effect of the nanowires is due to the connection of the n-type PC$_{61}$BM nanowires to the anode ITO/PEDOT:PSS layer, which increases the recombination rate between electrons and holes, resulting in a significant drop of short-circuit current density (J_{sc}) and fill factor (FF). In sharp contrast, the inverted cell with nanowire length (L_{NW}) = 120 nm and ρ_{NW} = 10^9 cm^{-2} nanowires exhibited an approximate 7 % increase in PCE compared with that without the nanowires (from 2.73 to 2.91 %), mainly thanks to the increase of J_{sc} (from 8.15 to 8.92 mA cm^{-2}). The presence of PC$_{61}$BM nanowires on the bottom cathode layer could merit electron collection, which leads to an improvement in the overall device performance.

Nevertheless, the area that the nanowires with L_{NW} = 120 nm and ρ_{NW} = 10^9 cm^{-2} can cover on the buffer layer is calculated as only 2 %. Therefore, the direct electron transport along the nanowires is limited, while the PCE improvement was much more. This implies that the nanowires act as seeds to promote the vertical segregation of PC$_{61}$BM during the formation of active layer. The role of SPNT fullerene nanowires is different from the cross-linked fullerene nanorods prepared by a template method [56]. The TEM images of BHJ with PC$_{61}$BM nanowires reveal that the nanowire seems to be larger than original nanowire radius, suggestive of the selective aggregation of PC$_{61}$BM onto the seed nanowires. However, a high density of nanowires ($\rho_{NW} > 10^{11}$ cm^{-2}) in turn caused a decrease of PCE. This is suggestive of a deformation of optimal BHJ network and/or damage of PC$_{61}$BM nanowires/buffer layer upon high-energy ion beams.

The LUMO of PC$_{61}$BM nanowire is a matter of great concern, because the LUMO levels are associated with the driving force of charge separation and V_{oc}. According to the density functional theory results of C$_{60}$ dimer [57, 58], the LUMOs are assumed slightly lowered in comparison with normal PC$_{61}$BM. In fact, V_{oc} of photovoltaic devices incorporating PC$_{61}$BM nanowires is decreased, and further decreased with increasing the number density and length of PC$_{61}$BM nanowires. On the other hand, V_{oc} does not change so much in the normal cells. This is consistent with the assumption that LUMO of PC$_{61}$BM nanowires became deeper.

Flash-photolysis time-resolved microwave conductivity (FP-TRMC) [59–61] is an electrode-less measurement tool of transient photoconductivity, which allows evaluation of nanometer-scale charge carrier mobility and charge separation yield in nanosecond time scale. In order to investigate the electronic property of

$PC_{61}BM$ nanowires, FP-TRMC experiments were performed for $PC_{61}BM$ films with/without ion beam exposure. The photoconductivity maxima were decreased by approximately 14 and 30 % for the films exposed to fluences of 10^{10} and 10^{11} cm^{-2}, respectively. Thus, the electron transport capability is degraded for the nanowires owing to the randomly polymerized network. This is consistent with the results for the OPV, in which the presence of excessive nanowires led to a significant decrease of PCE. It should be noted that this result is contrary to the report of a C_{60} film irradiated with 55–120 MeV ions, showing a formation of conductive pathways evaluated by conducting (c-) AFM [62]. Since the direction of charge transport (perpendicular for the c-AFM, parallel for FP-TRMC) and LET of ion beams are different, a direct comparison is not possible. The negative impact of ion beam exposure on the electronic device is more manifest in the normal $P3HT:PC_{61}BM$ cell incorporating P3HT nanowires ($L_{NW} = 200$ nm, $\rho_{NW} = 10^{11}$ cm^{-2}) instead of $PC_{61}BM$ nanowires. Despite the expected benefits of hole-conducting P3HT nanowires placed on the anode buffer layer, the PCE was extremely low, 0.53 % because of the low FF of 27.1 %.

The thermal stability of rigid fullerene nanowires should be emphasized and was examined. The AFM images of $PC_{61}BM$ nanowires did not change after thermal annealing at 160 °C for 10 min, demonstrating the feasibility of the rigid and covalently bonded nanowires as organic electronic component. In spite of the high mechanical strength, fullerene nanowires cannot stand on the surface after development. In a developer or active layer spin-coating, one end of the nanowires may be apart from the substrate; however, excessive long nanowires are expected not to keep the straight shape and finally knocked on the surface. The uncontrolled shape of nanowire might be one of the reasons for the decreased performance of OPV incorporating long nanowires. Making the nanowires stand on the surface is a great challenge to open an attractive avenue to the molecular electronics. Dry development process instead of the present *wet* process might be a key technique to this end.

The deployment of ICBA nanowires instead of $PC_{61}BM$ is a next step in OPV application, due to the promising high PCE of P3HT:PCBM. As introduced above, the radius of ICBA nanowires was 10.9 ± 1.1 nm, slightly larger than bare C_{60} (7.9 ± 1.0 nm) and PCBM (8.6 ± 0.8 nm), suggestive of the correlation between the density of the fullerene film and the nanowire radius [45]. Because ICBA, bulky indene bis-adduct of fullerene, is a mixture of stereoisomers which ensures the solubility in organic solvents by preventing the excessive crystallization, the film would have looser packing than mono-adduct PCBM and bare C_{60}.

BHJ-type OPVs of P3HT:ICBA were fabricated by incorporating ICBA nanowires with different lengths [63]. The thin ICBA films on $TiO_x/ITO/glass$ were exposed to 490 MeV Os ions at the fluence of 10^9 cm^{-2} and developed in chlorobenzene. This fluence was chosen on the basis of the successful example of P3HT:PCBM [45]. The P3HT:ICBA (1:1 w/w%) and PEDOT:PSS layers were subsequently spin-coated and annealed, and Au anode was thermally deposited in a vacuum chamber. The current density (J)–voltage (V) curves of these inverted cells under illumination from an AM 1.5 G solar simulator are shown in Fig. 6.7a.

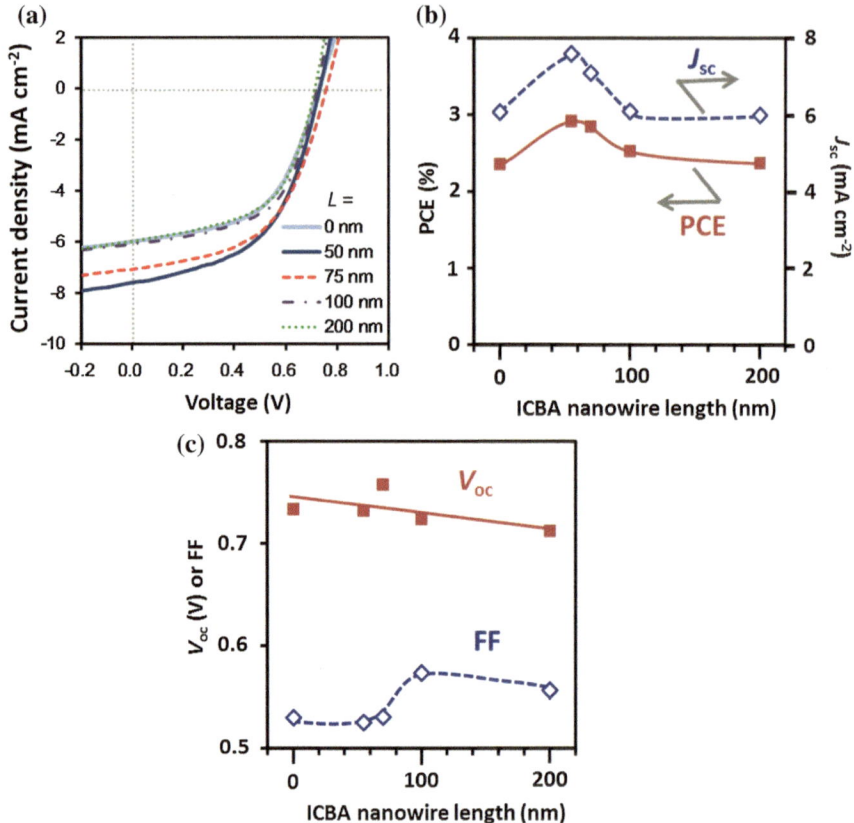

Fig. 6.7 a *J–V* curves of P3HT:ICBA (1:1 w/w) hybridized with ICBA nanowires $(1 \times 10^9 \text{ cm}^{-2})$ under AM 1.5 G (1 sun). *L* denotes the nanowire length. **b** PCE and J_{sc} versus nanowire length. **c** V_{oc} and FF versus nanowire length. Reproduced with permission from Ref. [63]. Copyright 2013 The Society of Photopolymer Science and Technology

The PCE, J_{sc}, V_{oc}, and FF are plotted in Fig. 6.7b, c, as a function of nanowire length (*L*). The $L = 0$ nm corresponds to the device without ICBA nanowires, but it was immersed in chlorobenzene prior to casting the active layer in the similar fashion with the hybridized devices. The PCE underwent 23 % enhancement from 2.36 to 2.92 % by incorporating 50-nm-length ICBA nanowires, mainly ascribed to the increase in J_{sc} (6.07–7.59 mA cm^{-2}). The favorable vertical segregation of *n*-type ICBA nanowires on the bottom cathode is assumed responsible for this improvement, being in line with the previous reports on the vertical segregation by post-annealing [64]. The V_{oc} (~0.73 V) and FF (0.53) are not affected so much for the short nanowires (<75 nm), and thus the PCEs of 50 and 75 nm nanowires were almost identical (2.92 and 2.84 %, respectively). These results look exactly like what we observed in OPV of PCBM nanowire-hybridized P3HT:PCBM [45], although the optimized length of nanowires became half of the PCBM case (ca. 120 nm). The observed V_{oc} is, however, lower than that of well-optimized normal cell of P3HT:ICBA without nanowires (0.87 V) [48], indicating that there are still margin of improvement by fine tuning of processing condition. The presence of lengthy nanowires (>200 nm) turned to deteriorate the device performance, in good correspondence with the trend found in PCBM nanowires [45]. The 200 nm length is comparable with the active layer thickness (ca. 220 nm), and thus ill-directed flow of charge carriers and/or trap at the electronically damaged parts of nanowires might be involved, leading to the increase of charge recombination that cannot contribute the photocurrent.

6.5 Nanowires for the Future Photo-Energy Conversion

The fullerene nanowires with thermally stable and high mechanical strength were successfully fabricated by SPNT. Notably, the nanowires are universally formed regardless of the cage size (C_{60} vs. C_{70}) and solubilizing substituents, consistent with the efficient 2+2 cycloaddition reaction induced by radiation. $PC_{61}BM$ and ICBA nanowires were hybridized with prototype P3HT-based OPV, demonstrating ca. 10–20 % improvement of PCE, mainly due to the increased J_{sc}. We have to note that the fullerene nanowire OPV is only beneficial for inverted cell structure, where the electron-conducting nanowires are placed on the bottom anode buffer layer. The nanowire length and density were found critical to the OPV performance, and a very small amount of nanowires is optimal for maximizing PCE, because excessive nanowires just impair the OPV output, due to the degraded electronic nature of fullerene nanowires by ion beam exposure. This work could open up opportunities to create a versatile platform of not only electronic applications but also manipulators, thereby widening the scope for potential application of this uniform, controllable, and semiconducting organic nanowires.

References

1. S.D. Dimitrov, J.R. Durrant, Chem. Mater. **26**, 616 (2014)
2. A.J. Heeger, Adv. Mater. **26**, 10 (2014)
3. L. Dou, J. You, Z. Hong, Z. Xu, G. Li, R.A. Street, Y. Yang, Adv. Mater. **25**, 6642 (2013)
4. R.A.J. Janssen, J. Nelson, Adv. Mater. **25**, 1847 (2013)
5. G. Li, R. Zhu, Y. Yang, Nat. Photon. **6**, 153 (2012)
6. F. He, L. Yu, J. Phys. Chem. Lett. **2**, 3102 (2011)
7. P.M. Beaujuge, J.M. Fréchet, J. Am. Chem. Soc. **133**, 20009 (2011)
8. P.-L.T. Boudreault, A. Najari, M. Leclerc, Chem. Mater. **23**, 456 (2011)
9. A.C. Arias, J.D. Mackenzie, I. McCulloch, J. Rivnary, A. Salleo, Chem. Rev. **110**, 3 (2010)
10. Y.J. Cheng, S.H. Yang, C.S. Hsu, Chem. Rev. **109**, 5868 (2009)
11. M.A. Green, K. Emery, Y. Hishikawa, W. Warta, E.D. Dunlop, Prog. Photovoltaics **23**, 1 (2015)
12. G. Yu, J. Gao, J.C. Hummelen, F. Wudl, A.J. Heeger, Science **270**, 1789–1791 (1992)

13. C.W. Isaacson, M. Kleber, J.A. Field, Environ. Sci. Technol. **43**, 6463–6474 (2009)
14. F. Giacalone, N. Martín, Chem. Rev. **106**, 5136–5190 (2006)
15. S.S. Babu, H. Mohwald, T. Nakanishi, Chem. Soc. Rev. **39**, 4021–4035 (2010)
16. Y. He, H.Y. Chen, J. Hou, Y. Li, J. Am. Chem. Soc. **132**, 1377–1382 (2010)
17. A.M. Nardes, A.J. Ferguson, J.B. Whitaker, B.W. Larson, R.E. Larsen, K. Maturová, P.A. Graf, O.V. Boltalina, S.H. Strauss, N. Kopidakis, Adv. Funct. Mater. **22**, 4115–4127 (2012)
18. C. Zhang, S. Chen, Z. Xiao, Q. Zuo, L. Ding, Org. Lett. **14**, 1508–1511 (2012)
19. K.-H. Kim, H. Kang, H.J. Kim, P.S. Kim, S.C. Yoon, B.J. Kim, Chem. Mater. **24**, 2373–2381 (2012)
20. H. Kang, C.-H. Cho, H.-H. Cho, T.E. Kang, H.J. Kim, K.-H. Kim, S.C. Yoon, B.J. Kim, ACS Appl. Mater. Interfaces **4**, 110–116 (2012)
21. Y. He, C. Chen, E. Richard, L. Dou, Y.G. Wu, Y. Lia, Y. Yang, J. Mater. Chem. **22**, 13391–13394 (2012)
22. S.A. Backer, K. Sivula, D.F. Kavulak, J.M.J. Fréchet, Chem. Mater. **19**, 2927–2929 (2007)
23. T. Mikie, A. Saeki, H. Masuda, N. Ikuma, K. Kokubo, S.J. Seki, Mater. Chem. A **3**, 1152–1157 (2015)
24. T. Mikie, A. Saeki, N. Ikuma, K. Kokubo, S. Seki, Chem. Lett. **44**, 282–284 (2015)
25. S.H. Park, A. Roy, S. Beaupré, S. Cho, N. Coates, J.S. Moon, D. Moses, M. Leclerc, K. Lee, A.J. Heeger, Nat. Photo. **3**, 297–303 (2009)
26. Z. He, C. Zhong, X. Huang, W.-Y. Wong, H. Wu, L. Chen, S. Su, Y. Cao, Adv. Mater. **23**, 4636–4643 (2011)
27. C. Guo, Y.-H. Lin, M.D. Witman, K.A. Smith, C. Wang, A. Hexemer, J. Strzalka, E.D. Gomez, R. Verduzco, Nano Lett. **13**, 2957–2963 (2013)
28. K. Nakabayashi, H. Mori, Macromolecules **45**, 9618–9625 (2012)
29. S. Miyanishi, Y. Zhang, K. Tajima, K. Hashimoto, Chem. Commun. **46**, 6723–6725 (2010)
30. N. Sary, F. Richard, C. Brochon, N. Leclerc, P. Leveque, J.N. Audinot, S. Berson, T. Heiser, G. Hadziioannou, R. Mezzenga, Adv. Mater. **22**, 763–768 (2010)
31. Q.L. Zhang, A. Cirpan, T.P. Russell, T. Emrick, Macromolecules **42**, 1079–1082 (2009)
32. Y.F. Tao, B. McCulloch, S. Kim, R.A. Segalman, Soft Matter **5**, 4219–4230 (2009)
33. D. Deribew, E. Pavlopoulou, G. Fleury, C. Nicolet, C. Renaud, S.-J. Mougnier, L. Vignau, E. Cloutet, C. Brochon, F. Cousin, G. Portale, M. Geoghegan, G. Hadziioannou, Macromolecules **46**, 3015–3024 (2013)
34. S. Moungthai, N. Mahadevapuram, P. Ruchhoeft, G. Stein, ACS Appl. Mater. Interfaces **4**, 4015–4023 (2012)
35. J. You, X. Li, F.-X. Xi, W.E.I. Sha, J.H.W. Kwong, G. Li, W.C.H. Choy, Y. Yang, Adv. Energy Mater. **2**, 1203–1207 (2012)
36. Y. Matsuo, Y. Sato, T. Niinomi, I. Soga, H. Tanaka, E. Nakamura, J. Am. Chem. Soc. **131**, 16048–16049 (2009)
37. X. Li, W.C.H. Choy, L. Huo, F. Xie, W.E.I. Sha, B. Ding, X. Guo, Y. Li, J. Hou, J. You, Y. Yang, Adv. Mater. **24**, 3046–3052 (2012)
38. V. Jankovic, Y.M. Yang, J. You, L. Dou, Y. Liu, P. Cheung, J.P. Chang, Y. Yang, ACS Nano **7**, 3815–3822 (2013)
39. C.-C. Chen, L. Dou, R. Zhu, C.-H. Chung, T.-B. Song, Y.B. Zheng, S. Hawks, G. Li, P.S. Weiss, Y. Yang, ACS Nano **6**, 7185–7190 (2012)
40. S.M. Yoon, S.J. Lou, S. Loser, J. Smith, L.X. Chen, A. Facchetti, T. Marks, Nano Lett. **12**, 6315–6321 (2012)
41. S. Seki, K. Maeda, S. Tagawa, H. Kudoh, M. Sugimoto, Y. Morita, H. Shibata, Adv. Mater. **13**, 1663–1665 (2001)
42. S. Seki, A. Saeki, W. Choi, Y. Maeyoshi, A. Omichi, A. Asano, K. Enomoto, C. Vijayakumar, M. Sugimoto, S. Tsukuda, S. Tanaka, J. Phys. Chem. B **116**, 12857–12863 (2012)
43. Cheng, H.L., Tang, M.T., Tuchinda, W., Enomoto, K., Chiba, A., Saito, Y., Kamiya, T., Sugimoto, M., Saeki, A., Sakurai, T., Omichi, M., Sakamaki, D., Seki, S.: Adv. Mater. Interfaces **2**, 1400450/1–9 (2015)

44. S. Tsukuda, S. Seki, S. Tagawa, M. Sugimoto, A. Idesaki, S. Tanaka, A. Oshima, Fabrication of nanowires using high-energy ion beams. J. Phys. Chem. B **108**, 3407–3409 (2004)

45. Maeyoshi, Y., Saeki, A., Suwa, S., Omichi, M., Marui, H., Asano, A., Tsukuda, S., Sugimoto, M., Kishimura, A., Kataoka, K., Seki, S.: Sci. Rep. **2**, 600/1–6 (2012)

46. H.W. Kroto, J.R. Heath, S.C. O'Brien, R.F. Curl, R.E. Smalley, Nature **318**, 162–163 (1985)

47. J.C. Hummelen, B.W. Knight, F. LePeq, F. Wudl, J. Yao, C.L. Wilkins, J. Org. Chem. **60**, 532–538 (1995)

48. G. Zhao, Y. He, Y. Li, Adv. Mater. **22**, 4355–4358 (2010)

49. A.M. Rao, P. Zhou, K.-A. Wang, G.T. Hager, J.M. Holden, Y. Wang, W.-T. Lee, X.-X. Bi, P.C. Eklund, D.S. Cornett, M.A. Duncan, J. Amster, Science **259**, 955–957 (1993)

50. P. Zhou, Z.-H. Dong, A.M. Rao, P.C. Eklund, Chem. Phys. Lett. **211**, 337–340 (1993)

51. Y. Iwasa, T. Arima, R.M. Fleming, T. Siegrist, O. Zhou, R.C. Haddon, L.J. Rothberg, K.B. Lyons, H.L. CarterJr, A.F. Hebard, R. Tycko, G. Dabbagh, J.J. Krajewski, G.A. Thomas, T. Yagi, Science **264**, 1570–1572 (1994)

52. S. Malik, N. Fujita, P. Mukhopadhyay, Y. Goto, K. Kaneko, T. Ikeda, S. Shinkai, J. Mater. Chem. **17**, 2454–2458 (2007)

53. Seki, S., Tsukuda, S., Maeda, K., Matsui, Y., Saeki, A., Tagawa, S.: Phys. Rev. B **70**, 144203/1–8 (2004)

54. W. Burlant, J. Neerman, V. Serment, J. Polym. Sci. **58**, 491–500 (1962)

55. W. Zhang, W. Jin, T. Fukushima, A. Saeki, S. Seki, T. Aida, Science **334**, 340–343 (2011)

56. C.-Y. Chang, C.-E. Wu, S.-Y. Chen, C. Cui, Y.-J. Cheng, C.-S. Hsu, Y.-L. Wang, Y. Li, Angew. Chem. Int. Ed. **123**, 9558–9562 (2011)

57. A. Bihlmeier, C.C.M. Samson, W. Klopper, ChemPhysChem **6**, 2625–2632 (2005)

58. T. Ren, B. Sun, Z. Chen, L. Qu, H. Yuan, X. Gao, S. Wang, R. He, R. Zhao, Y. Zhao, Z. Liu, X. Jing, J. Phys. Chem. B **111**, 6344–6348 (2007)

59. A. Saeki, Y. Koizumi, T. Aida, S. Seki, Acc. Chem. Res. **45**, 1193–1202 (2012)

60. T.J. Savenije, D.H.K. Murthy, M. Gunz, J. Gorenflot, L.D.A. Siebbeles, V. Dyakonov, C. Deibel, J. Phys. Chem. Lett. **2**, 1368–1371 (2011)

61. F.C. Grozema, L.D.A. Siebbeles, J. Phys. Chem. Lett. **2**, 2951–2958 (2011)

62. Kumar, A., Avasthi, D. K., Tripathi, A., Kabiraj, D., Singh, F., Pivin J.C.: J. Appl. Phys. **101**, 014308/1–5 (2007)

63. S. Suwa, Y. Maeyoshi, S. Tsukuda, M. Sugimoto, A. Saeki, S.J. Seki, Photopolym. Sci. Tech. **26**, 193–197 (2013)

64. Orimo, A., Masuda, K., Honda, S., Benten, H., Ito, S., Ohkita, H., Tsuji, H.: Appl. Phys. Lett. **96**, 043305/1–3 (2010)

Chapter 7
Single-Particle Triggered Polymerization

7.1 Preliminary Remarks

Well-defined nano-sized objects [1, 2] have attracted increasing attentions due to the recent advances in materials science and nanotechnology. In particular, one-dimensional (1D) nanostructures from π-conjugated molecules serve as promising materials for optical and electronic applications [3–8] such as chemical sensors, memory devices, organic photovoltaic cells (OPVC), and organic field-effect transistors (OFET). Because organic materials are soft and finely tunable by molecular design, most of these nanostructures are constructed by bottom-up self-assembly approaches with non-covalent interactions of the elaborately designed π-conjugated molecules or polymers [9–11]. On the other hand, top-down technologies to access such 1D nanostructures are attractive in view of generality and versatility but have been still limited [12–14]. Seki et al. have reported a unique top-down technology using swift heavy ions, 1D nanostructures were developed by employing cross-linking reactions of macromolecules in the film state. This method is named "single-particle nanofabrication technique (SPNT)" [15]. In SPNT, the MeV-order swift heavy ions, irradiated to the target polymer films non-homogeneously, produce the reactive intermediates—ion radicals, neutral radicals, etc.—along their trajectories called ion tracks. Because of effective and high-density formation of neutral radical species generated in radiation cross-linking type polymers [16, 17], they induce the recombination reactions within the ion track, forming wire-shaped insoluble nanogels with fairly controlled sizes. The obtained nanowires are isolated by development process, where the non-irradiated area in the film is washed away by organic solvents (Fig. 7.1). Although SPNT has been utilized so far for various types of synthetic polymers [18–21] and biomacromolecules [22], however, there is only one example that nanowires are available from small molecular materials by analogous means of SPNT [23], where high reactivity of fullerene derivatives upon ion beam irradiation [24] was utilized.

© The Author(s) 2015
S. Seki et al., *High-Energy Charged Particles*,
SpringerBriefs in Molecular Science, DOI 10.1007/978-4-431-55684-8_7

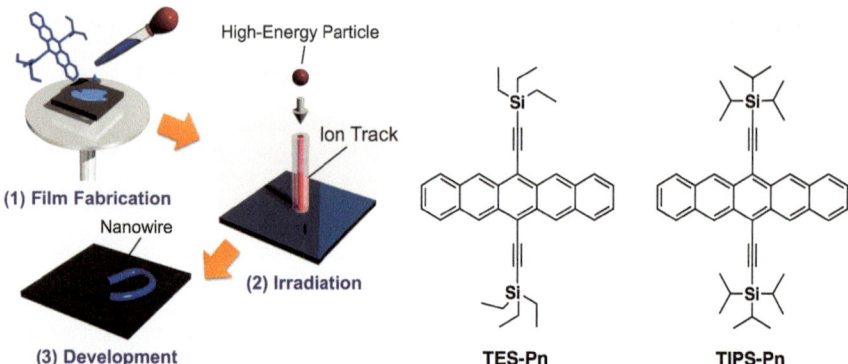

Fig. 7.1 Schematic illustrations of nanowire fabrication processes and chemical structures of soluble pentacene derivatives, 6,13-bis((triethylsilyl)ethynyl)pentacene (TES-Pn), and 6,13-bis((triisopropylsilyl)ethynyl)pentacene (TIPS-Pn). Reprinted with permission from Ref. [24] ©2015, VBRI press

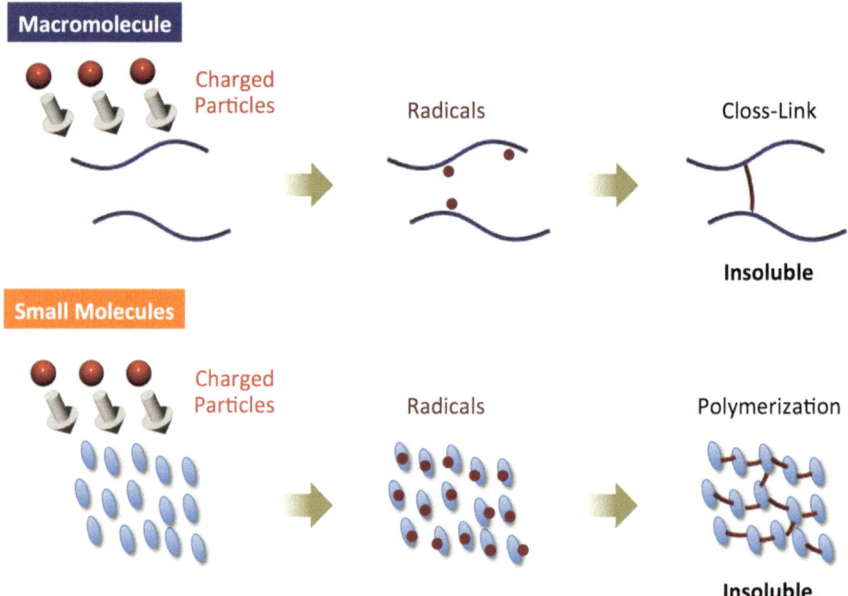

Fig. 7.2 Schematic illustrations of insoluble nanogel formation by chemical reaction of high-energy particles. For small molecules, polymerization reaction is necessary to yield nanogels

In principle, nanowire formation by single-particle irradiation would be more difficult for small molecules because it requires a larger number of reaction points than the case with macromolecules (Fig. 7.2). Toward this background, this chapter reviews the recent results featuring the development of organic nanowires from small molecular organic compounds through intra-track chemical reactions using

ion beams [25]. Thin films of pentacene derivatives, 6,13-bis(triethylsilylethynyl)
pentacene (TES-Pn), and 6,13-bis((triisopropylsilyl)ethynyl)pentacene (TIPS-Pn)
(Fig. 7.1) were subjected to high-energy particle irradiation at the fluence of 10^8–
10^{10} cm^{-2} and thereafter developed by organic solvents. This method, referred to
as Single-particle Triggered Linear Polymerization (STLiP), afforded the isolation
of wire-shaped nanomaterials on a substrate that was visualized by atomic force
microscopy and scanning electron microscopy.

7.2 Formation of Nanowires from Alkyne-Functionalized Compounds

Pentacene is one of the most intensively studied organic semiconducting motifs
and often used as an OFET active layer because of its high hole mobility [26, 27].
Nevertheless, pentacene is sensitive to oxygen and in fact unstable under ambient
condition. To improve the stability and solubility of pentacene, silylethyne-substituted pentacenes were reported and proved their excellent carrier transporting performance [28, 29]. Considering the easy isolation by the development procedure in
STLiP, thin films of soluble pentacenes, 6,13-bis((trialkylsilyl)ethynyl)pentacenes
were fabricated and developed after irradiation of swift heavy ions using organic
solvent.

When the soluble pentacene films were exposed to 490 MeV Os particles at
the fluence of 5.0×10^8–1.0×10^9 ions cm^{-2} followed by the development in
hexane, nanowire objects were visualized in AFM (Fig. 7.3). By carefully looking
at the images, partial fragmentation of the nanowires was confirmed. Nevertheless,
even for the thick (>1 μm) films prepared by drop-cast method, nanowires with
ultrahigh aspect ratio were observed. Compared to the previous reports of SPNT-
based nanowires [18–23], the obtained nanowires look flexible and their diameters appear small. Due to such flexibility, network structures of nanowires may be
formed, which was more obviously observed in SEM. It is considered that these
network structures were resulted from the entanglement of the formed nanowires
one another. In contrast, irradiation to pristine pentacene films did not afford any
nanowire. These results indicate that silylethyne moieties in the pentacene derivatives give the high reaction efficiency and/or significant difference in solubility. In
regard to this possibility, Seki et al. have already reported that introduction of ter-
minal alkyne groups into the polystyrene backbone provides efficient cross-linking
reactions in SPNT [21]. Therefore, it was considered that alkyne groups would
increase the reactivity upon swift heavy ion irradiations.

The radius of nanowire was evaluated by the cross-sectional profile in AFM.
After the development process, the cross section of nanowires on the substrate was
elliptically deformed as a consequence of adsorption forces. Taking into account
this issue, the radius of nanowire is calculated by applying the ellipse model to
cross section of nanowire (Fig. 7.4d). The values of r_x and r_y are defined as the
half-width and half-height at half-maximum of the cross section of nanowires,

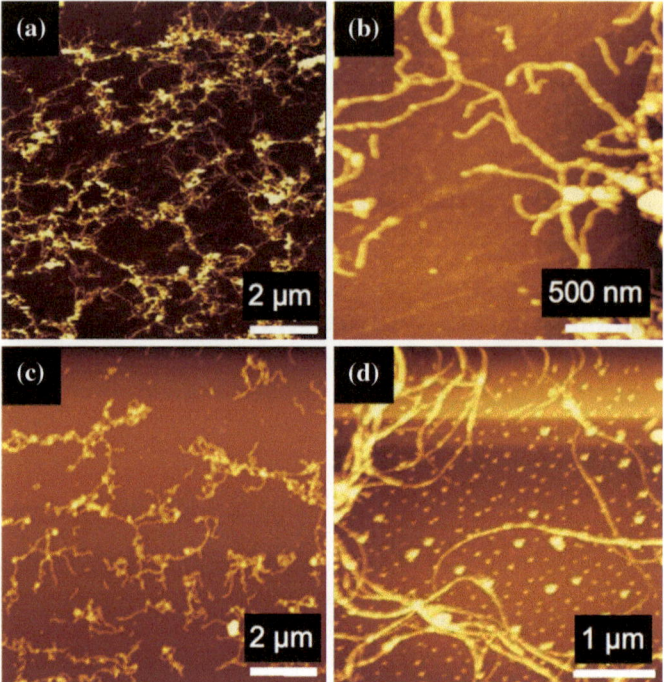

Fig. 7.3 AFM topographic images of nanowires isolated via development of **a, b** TES-Pn and **c, d** TIPS-Pn drop-cast films with *n*-hexane after irradiation with 490 MeV ^{192}Os^{30+} particles at the fluence of 1.0×10^9 and 5.0×10^8 ions cm^{-2} for (**a**) and (**b–d**), respectively. Reprinted with permission from Ref. [24] ©2015, VBRI press

respectively. Figure 7.4c shows the cross-sectional profile of the nanowire based on TES-Pn. The profile was measured on the line in Fig. 7.4b. The radii of the nanowires from TES-Pn and TIPS-Pn are calculated as 4.5 and 4.2 nm, respectively.

In order to investigate the reactivity of silylethyne-substituted pentacene, the overall propagation/cross-linking efficiency G was evaluated. The value of G is defined as the number of reactions of the resulting polymer propagation and/ or cross-linking in the isolated nanostructures per 100 eV of absorbed dose. Considering the energy distribution in an ion track, the size of the nanowire was well interpreted by the theoretical model. Again G is given by Eq. (6.1), and the extremely high G values for TES- and TIPS-pentacene of $G_{\text{TES-Pn}} > 7$ and $G_{\text{TIPS-Pn}} > 5$ (100 eV)$^{-1}$ were evaluated. In general, deprotection of trialkylsilyl groups is performed with acids, bases, or fluorides. The relative stability of trialkylsilyl protection can be tuned by changing the volume of substituents [30]. TIPS group indicates higher resistance for these conditions than TES group. Therefore, it is expected that TIPS group controls the generation of radical intermediates and promotion of the polymerization/cross-linking reactions.

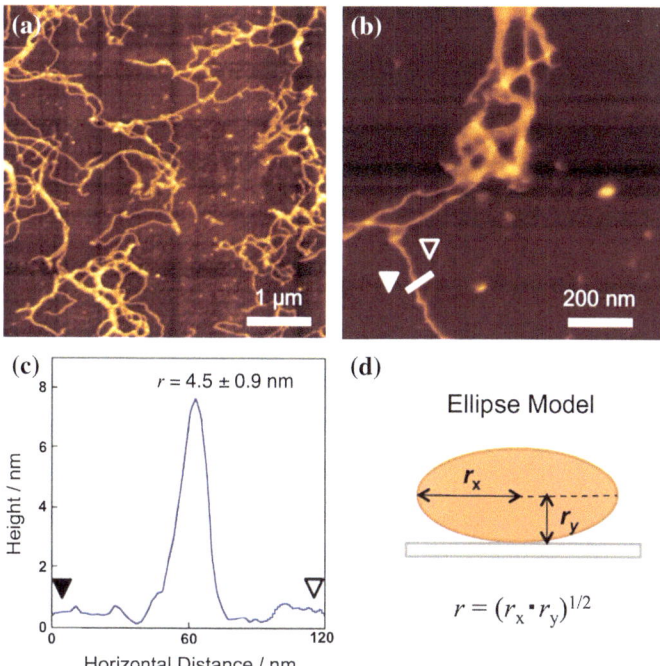

Fig. 7.4 a, b AFM topographic images of nanowires isolated via development of 1.7-μm-thick TES-Pn film with *n*-hexane after irradiation with 490 MeV $^{192}Os^{30+}$ particles at the fluence of 5.0×10^8 ions cm^{-2}. **c** Cross-sectional profiles of the nanowires on Si substrate based on TES-pentacene produced by STLiP. The profile was measured at the position indicated by lines in (**b**). **d** Schematic illustration of nanowire radius using ellipse model. Reprinted with permission from Ref. [24] ©2015, VBRI press

7.3 High-Energy Charged Single-Particle as a Versatile Tool for Nanofabrication of Organic Materials ~From Cross-linking to Polymerization~

The new concept of nanofabrication classified as Single-particle Triggered Linear Polymerization (STLiP) is introduced in the present chapter, demonstrated as the direct formation of nanowires from pentacene derivatives. This one-dimensional gelation event was triggered by the effective initiation/propagation reaction of intra-track polymerization as well as cross-linking along the trajectories of incident high-energy charged particles. Although the size of molecules is much smaller than polymeric materials, the pentacene derivatives carrying silylethyne groups gave flexible nanowires in STLiP that was strongly entangled into network structures. Based on the observations, the silylethyne groups provide appropriate solubility and increased overall reaction efficiencies of propagation and cross-linking reactions upon irradiation of swift heavy ions. Considering the prominent semiconducting property of pentacene and observed unique network structures, together with potential universality of the present technique to other small molecular systems, one can believe that nanowires by STLiP encourage the development of future electronic devices.

References

1. C. Sanchez, P. Belleville, M. Popall, L. Nicole, Chem. Soc. Rev. **40**, 696 (2011)
2. Y.-Z. Long, M. Yu, B. Sun, C.-Z. Gu, Z. Fan, Chem. Soc. Rev. **41**, 4560 (2012)
3. Y.S. Zhao, H. Fu, A. Peng, Y. Ma, D. Xiao, J. Yao, Adv. Mater. **20**, 2859 (2008)
4. Y.S. Zhao, H.B. Fu, F.Q. Hu, A.D. Peng, W.S. Yang, J.N. Yao, Adv. Mater. **20**, 79 (2008)
5. P.G. Bruce, B. Scrosati, J.-M. Tarascon, Angew. Chem. Int. Ed. **47**, 2930 (2008)
6. F.S. Kim, G. Ren, S.A. Jenekhe, Chem. Mater. **23**, 682 (2011)
7. X. Chen, C.K.Y. Wong, C.A. Yuan, G. Zhang, Sensors Actuators B: Chem. **177**, 178 (2013)
8. A.I. Hochbaum, P. Yang, Chem. Rev. **110**, 527 (2010)
9. L. Zang, Y. Che, J.S. Moore, Acc. Chem. Res. **41**, 1596 (2008)
10. C.C. Lee, C. Grenier, E.W. Meijer, A.P.H.J. Schenning, Chem. Soc. Rev. **38**, 671 (2009)
11. C. Giansante, G. Raffy, C. Schäfer, H. Rahma, M.-T. Kao, A.G.L. Olive, A. Del Guerzo, J. Am. Chem. Soc. **133**, 316 (2011)
12. A. Greiner, J.H. Wendorff, Angew. Chem. Int. Ed. **46**, 5670 (2007)
13. H. Cho, S.-Y. Min, T.-W. Lee, Macromol. Mater. Eng. **298**, 475 (2013)
14. S.-Y. Min, T.-S. Kim, B.J. Kim, H. Cho, Y.-Y. Noh, H. Yang, J.H. Cho, T.-W. Lee, Nat. Commun. **4**, 1773 (2013)
15. S. Seki, K. Maeda, S. Tagawa, H. Kudoh, M. Sugimoto, Y. Morita, H. Shibata, Adv. Mater. **13**, 1663 (2001)
16. A.A. Miller, E.J. Lawton, J.S. Balwit, J. Poly. Sci. **14**, 503 (1954)
17. J.K. Thomas, Nucl. Instr. Meth. B **265**, 1 (2007)
18. S. Seki, S. Tsukuda, K. Maeda, S. Tagawa, H. Shibata, M. Sugimoto, K. Jimbo, I. Hashitomi, A. Kohyama, Macromolecules **38**, 10164 (2005)
19. S. Tsukuda, S. Seki, M. Sugimoto, S. Tagawa, J. Phys. Chem. B **110**, 19319 (2006)
20. S. Seki, A. Saeki, W. Choi, Y. Maeyoshi, A. Omichi, A. Asano, K. Enomoto, C. Vijayakumar, M. Sugimoto, S. Tsukuda, S. Tanaka, J. Phys. Chem. B **116**, 12857 (2012)
21. A. Asano, M. Omichi, S. Tsukuda, K. Takano, M. Sugimoto, A. Saeki, S. Seki, J. Phys. Chem. C **116**, 17274 (2012)
22. M. Omichi, A. Asano, S. Tsukuda, K. Takano, M. Sugimoto, A. Saeki, D. Sakamaki, A. Onoda, T. Hayashi, S. Seki, Nat. Commun. **5**, 3718 (2014)
23. Y. Maeyoshi, A. Saeki, S. Suwa, M. Omichi, H. Marui, A. Asano, S. Tsukuda, M. Sugimoto, A. Kishimura, K. Kataoka, S. Seki, Sci. Rep. **2**, 600 (2012)
24. A. Kumar, D.K. Avasthi, A. Tripathi, D. Kabiraj, F. Singh, J.C. Pivin, J. Appl. Phys. **101**, 014308 (2007)
25. Y. Takeshita, T. Sakurai, A. Asano, K. Takano, M. Omichi, M. Sugimoto, S. Seki, Adv. Mat. Lett. **6**, 99–103 (2015)
26. V.Y. Butko, X. Chi, D.V. Lang, A.P. Ramirez, Appl. Phys. Lett. **83**, 4773 (2003)
27. T.-H. Chao, M.-J. Chang, M. Watanabe, M.-H. Luo, Y.J. Chang, T.-C. Fang, K.-Y. Chen, T.J. Chow, Chem. Commun. **48**, 6148 (2012)
28. J.E. Anthony, J.S. Brooks, D.L. Eaton, S.R. Parkin, J. Am. Chem. Soc. **123**, 9482 (2001)
29. M.M. Payne, S.R. Parkin, J.E. Anthony, C.-C. Kuo, T.N. Jackson, J. Am. Chem. Soc. **127**, 4986 (2005)
30. Nelson, T.D., Crouch, R.D.: Synthesis 1031 (1996)